METHODS IN MOLECULAR BIOLOGY™

Series Editor
John M. Walker
School of Life Sciences
University of Hertfordshire
Hatfield, Hertfordshire, AL10 9AB, UK

For other titles published in this series, go to
www.springer.com/series/7651

RNAi and MicroRNA-Mediated Gene Regulation in Stem Cells

Methods, Protocols, and Applications

Edited by

Baohong Zhang and Edmund J. Stellwag

Department of Biology, East Carolina University, Greenville, NC, USA

Humana Press

Editors
Baohong Zhang
Department of Biology
East Carolina University
27858-4353 Greenville North Carolina
Howell Science Complex
USA
zhangb@ecu.edu

Edmund J. Stellwag
Department of Biology
East Carolina University
27858-4353 Greenville North Carolina
Howell Science Complex
USA
stellwage@ecu.edu

ISSN 1064-3745 e-ISSN 1940-6029
ISBN 978-1-4939-5749-1 ISBN 978-1-60761-769-3 (eBook)
DOI 10.1007/978-1-60761-769-3
Springer New York Dordrecht Heidelberg London

Cover illustration: Lentiviral transduction efficiency in hMSCs

Printed on acid-free paper

Humana Press is part of Springer Science+Business Media (www.springer.com)

Preface

At present, research in the areas of stem-cell biology, RNA interference (RNAi), and microRNAs (miRNAs) is the source of immense scientific and medical interest. Despite their apparent differences, stem-cell biology, RNAi, and miRNA are increasingly being shown to have a great deal in common. Small RNAs, particularly those represented by miRNAs and RNAi, play a central role in the regulation of embryonic and adult stem-cell development. Recent research has revealed that in stem cells, miRNA- and RNAi-mediated gene regulation is one of the determinates controlling the state of cell differentiation, with the small RNAs serving as key elements involved in regulatory network control of pluripotent cell-fate determination. From a technological perspective, RNAi and miRNA-based methods are being developed both to study the processes of stem-cell differentiation and create clinically useful therapies for the treatment of disease and preservation of wellness. The promise is that harnessing the stem-cell regulatory activities of RNAi and miRNA will lead to rapid advancement in the use of these small RNA molecules as therapeutic agents in areas of medicine related to stem-cell biology, including regenerative medicine, aging, cancer, and neurological disorders, to name a few.

When the present book project was undertaken, the goal was to provide an accessible compendium of up-to-date methods focused on the study of RNAi and miRNAs-mediated gene regulation in stem cells. We believed this was a noteworthy goal because the fields of RNAi, miRNA, and stem-cell biology have been undergoing explosive growth, which makes it difficult for anyone to keep abreast of the seminal methods used to conduct experiments and analyze results. We recognized that the availability of detailed, user-tested protocols for conducting RNAi and miRNA-based research on stem cells would accelerate progress in the field by reducing the time required to decipher and put into practice procedures published in the literature.

In order to achieve our goal, we realized that this effort would be beyond the scope of a single individual or laboratory. We, therefore, enlisted the help of a team of knowledgeable experts from academia and industry to assist by contributing chapters that would cover the various methodological approaches presently being used to examine RNAi and miRNA in stem cells. As a testament to their expertise, the majority of the chapters are devoted to methods that have been developed in the authors' laboratories and adopted by other research laboratories around the world. We feel fortunate that the authors have agreed to provide detailed protocols, including information about the source of reagents, cell lines, and other specialized items used in the protocols. We are also pleased that the authors have made a special effort to share their insights and strategic viewpoints on all three subjects: RNAi, miRNAs, and stem cells.

The book is divided into three major sections. The first two chapters give a brief introduction to RNAi and miRNAs in stem cells, with a focus on the current status of research and future perspectives. The second section is focused on the methods and protocols for RNAi screening, transfection, and the knockdown of specific genes and pathways in several animal species, including humans and mice. The third section centers on recently developed methods for identification of miRNAs and includes a general protocol for preparation and analysis of miRNA libraries for deep sequencing, knockdown of a specific gene using miRNA-based shRNA, and miRNA expression analysis using qRT-PCR.

We want to take this opportunity to thank all the authors, who have contributed excellent chapters to this book. Their special effort to provide detailed protocols makes the book a valuable resource for scientists and aspiring graduate students interested in the intersection of RNAi, miRNA, and stem-cell molecular biology. We also want to express our sincere appreciation to Professor John M. Walker, the Methods in Molecular Biology Series Editor, and Mr. David Casey from Humana Press, for their invitation to edit this book and help, support, and commitment during its preparation.

Baohong Zhang
Edmund J. Stellwag

Contents

Contributors

AURÉLIE BAUDET • *Molecular Medicine and Gene Therapy, Lund Stem Cell Center, Lund University, Lund, Sweden*

SOIZIK BERLIVET • *Epigenetic Regulation and Cancer Group, CEA, iBiTecS, Gif-sur-Yvette Cedex, France; Department of Obstetrics and Gynecology, McGill University, Montréal, QC, Canada*

DETLEV BOISON • *Robert S. Dow Neurobiology Laboratories, Legacy Research, Portland, OR, USA*

KIN CHAN • *Department of Biochemistry, Institute for Stem Cell and Regenerative Medicine, University of Washington, Seattle, WA, USA*

TZU-CHING CHANG • *Institute of Cellular and System Medicine, National Health Research Institutes, Zhunan, Miaoli County, Taiwan*

STEPHEN W. CHEN • *Stem Cell Group, Bioprocessing Technology Institute, Centros, Singapore; 2032 Newport Way NW, Issaquah, WA 98027, USA*

XIAOXING CHENG • *Division of Research, Beijing 309 Hospital, 17 Heishanhu Street, Haidian, Beijing 100091, China*

NOREEN DHALLA • *Genome Sciences Centre, BC Cancer Agency, Vancouver, BC, Canada*

MICHAEL FLENTJE • *Department of Radiation Oncology, University of Würzburg, Würzburg, Germany*

YUMEI FU • *Department of Pathology and Otolaryngology, Johns Hopkins University School of Medicine, Baltimore, MD, USA*

MATTHIEU GÉRARD • *Epigenetic Regulation and Cancer Group, CEA, iBiTecS, Gif-sur-Yvette Cedex, France*

CARSTEN HAGEMANN • *Tumorbiology Laboratory, Department of Neurosurgery, University of Würzburg, Würzburg, Germany*

MARTIN HIRST • *Genome Sciences Centre, BC Cancer Agency, Vancouver, BC, Canada*

MARTIN HOULARD • *Epigenetic Regulation and Cancer Group, CEA, iBiTecS, Gif-sur-Yvette Cedex, France; Institute Pasteur, Paris, France*

RUI JIAN • *Division of Research, Beijing 309 Hospital, 17 Heishanhu Street, Haidian, Beijing 100091, China*

CHRISTINE KARLSSON • *Molecular Medicine and Gene Therapy, Lund Stem Cell Center, Lund University, Lund, Sweden*

BOR-SHENG KO • *Institute of Cellular and System Medicine, National Health Research Institutes, Zhunan, Miaoli County, Taiwan; Department of Internal Medicine, National Taiwan University Hospital, Taipei, Taiwan*

JONAS LARSSON • *Molecular Medicine and Gene Therapy, Lund Stem Cell Center, Lund University, Lund, Sweden*

CYNTHIA LEBRON • *Department of Pathology and Otolaryngology, Johns Hopkins University School of Medicine, Baltimore, MD, USA*

JUN-YANG LIOU • *Institute of Cellular and System Medicine, National Health Research Institutes, Zhunan, Miaoli County, Taiwan*

WEI LU • *Laboratory of Molecular Neural Biology, School of Life Sciences, Shanghai University, Shanghai, China*

TAKAYUKI MANABE • *Department of 2nd Anatomy, Faculty of Medicine, Nara Medical University, Nara, Japan*

MARCO MARRA • *Genome Sciences Centre, BC Cancer Agency, Vancouver, BC, Canada*

HIDEAKI MATSUOKA • *Department of Biotechnology and Life Science, Tokyo University of Agriculture and Technology, Tokyo, Japan*

HELEN MCDONALD • *Genome Sciences Centre, BC Cancer Agency, Vancouver, BC, Canada*

RYAN D. MORIN • *Genome Sciences Centre, BC Cancer Agency, Vancouver, BC, Canada*

ALEXANDER K. MURASHOV • *Department of Physiology, East Carolina University School of Medicine, Greenville, NC, USA*

STEVE K.W. OH • *Stem Cell Group, Bioprocessing Technology Institute, Centros, Singapore*

XIAOPING PAN • *Department of Biology, East Carolina University, Greenville, NC, USA*

PAWAN PANDOH • *Genome Sciences Centre, BC Cancer Agency, Vancouver, BC, Canada*

ANNA-LIISA PRABHU • *Genome Sciences Centre, BC Cancer Agency, Vancouver, BC, Canada*

GAOYING REN • *Department of Medicine, University of Washington, Seattle, WA, USA*

KLAUS ROOSEN • *Tumorbiology Laboratory, Department of Neurosurgery, University of Würzburg, Würzburg, Germany*

HANNELE RUOHOLA-BAKER • *Department of Biochemistry, Institute for Stem Cell and Regenerative Medicine, University of Washington, Seattle, WA, USA*

HARUN M. SAID • *Department of Radiation Oncology, University of Würzburg, Würzburg, Germany*

MIKAKO SAITO • *Department of Biotechnology and Life Science, Tokyo University of Agriculture and Technology, Tokyo, Japan*

EDMUND J. STELLWAG • *Department of Biology, East Carolina University, Greenville, NC, USA*

ANGELA TAM • *Genome Sciences Centre, BC Cancer Agency, Vancouver, BC, Canada*

MELANIE I. TROMBLY • *Department of Biochemistry, Molecular Biology, and Cell Biology, Northwestern University, Evanston, IL, USA*

GILES HAMILTON VINCE • *Tumorbiology Laboratory, Department of Neurosurgery, University of Würzburg, Würzburg, Germany*

AKIO WANAKA • *Department of 2nd Anatomy, Faculty of Medicine, Nara Medical University, Nara, Japan*

JIANLONG WANG • *Department of Developmental and Regenerative Biology, Black Family Stem Cell Institute, Mount Sinai School of Medicine, New York, NY, USA*

JIAO WANG • *Laboratory of Molecular Neural Biology, School of Life Sciences; Institute of Systems Biology, Shanghai University, Shanghai, China*

XIAOZHONG WANG • *Department of Biochemistry, Molecular Biology, and Cell Biology, Northwestern University, Evanston, IL, USA*

TIEQIAO WEN • *Laboratory of Molecular Neural Biology, School of Life Sciences; Institute of Systems Biology, Shanghai University, Shanghai, China*

THOMAS ZENG • *Genome Sciences Centre, BC Cancer Agency, Vancouver, BC, Canada*

BAOHONG ZHANG • *Department of Biology, East Carolina University, Greenville, NC, USA*

YONGJUN ZHAO • *Genome Sciences Centre, BC Cancer Agency, Vancouver, BC, Canada*

GANG-MING ZOU • *Department of Pathology and Otolaryngology, Johns Hopkins University School of Medicine, Baltimore, MD, USA; State Key Laboratory for Oncogene and Cancer-Associated Gene, Shanghai Cancer Institute, Shanghai Jiaotong University, 800 Dong Chuan Road, Shanghai 200240, People's Republic of China*

Section I

Introduction

Chapter 1

RNAi in Stem Cells: Current Status and Future Perspectives

Gang-Ming Zou

Abstract

RNAi is a mechanism displayed by most eukaryotic cells to rid themselves of foreign double-strand RNA molecules. In the 11 years since the initial report, RNAi has now been demonstrated to function in mammalian cells to alter gene expression and used as a means for genetic discovery as well as a possible strategy for genetic correction and genetic therapy in cancer and other diseases. The aim of this review is to provide a general overview of how RNAi suppresses gene expression, examine some published RNAi approaches that have resulted in changes in stem cell function, and suggest the possible clinical relevance of this work in cancer therapy through targeting cancer stem cells.

Key words: RNAi, siRNA, ES cells, stem cells, cancer stem cells, Shp-2, Ape1.

Abbreviations

RNAi	RNA interference
siRNA	small interfering RNA
dsRNA	double-strand RNA
miRNA	microRNA
PGC	primordial germ cell
PKR	dsRNA-dependent protein kinase
shRNA	small hairpin RNA
HSC	hematopoietic stem cells
AAV	adenoviral-associated vector
RISC	siRNA-induced silencing complex

1. Introduction

RNAi was defined by Fire and his colleagues (1) to describe the inhibition of gene expression by double-strand RNAs (dsRNAs) when introduced into worms. Following on the studies of Guo and Kemphues (2), who had found that sense RNA was as effective as antisense RNA for suppressing gene expression in

B. Zhang, E.J. Stellwag (eds.), *RNAi and MicroRNA-Mediated Gene Regulation in Stem Cells*, Methods in Molecular Biology 650, DOI 10.1007/978-1-60761-769-3_1, © Springer Science+Business Media, LLC 2010

worms, Fire and his colleagues (1) applied single-strand antisense RNA and double-strand RNA in their experiments. To their surprise, they found that double-strand RNA was more effective at producing interference than was either strand individually. After injection into adult *Caenorhabditis elegans*, single-strand antisense RNA had a modest effect in diminishing specific gene expression whereas double-stranded mixtures caused potent and specific interference. Because of his outstanding achievement, Fire was awarded the Nobel Prize in Physiology and Medicine in 2006. RNAi is a multistep process involving the generation of small interfering RNAs (siRNAs) in vivo through the action of the RNase III endonuclease Dicer. The resulting 21- to 23-nt siRNAs mediate degradation of their complementary RNA (3). The following sections will review the basic mechanisms of RNAi, introduce some studies demonstrating the usefulness of this strategy to modulate gene expression in mammalian cells, and describe potential uses of RNAi to alter stem cell functions, including cancer stem cells.

2. Basic Aspects of RNAi

It has been suggested that long dsRNAs are the precursors of siRNAs that can trigger RNAi. When dsRNAs enter into cells, they are cleaved by an RNase III like enzyme, Dicer, into siRNAs. These newly formed 21- to 23-nt siRNA molecules bind to several proteins to form an siRNA–protein complex, which contains a helicase activity that unwinds the two strands of RNA molecules allowing the antisense strand to bind the targeted RNA molecules (4, 5) and an endonuclease activity that hydrolyzes the target RNA at the site where the antisense strand is bound. Formation of the siRNA-induced silencing complex (RISC) is the key process in siRNA-directed mRNA degradation. The RISC complex includes helicase, or exonuclease and a PAZ/Piwi protein{Qcer: please chk comment if ok.} (rde1) (6, 7). The RISC complex is responsible for the sequence-specific degradation of the target RNAs that contain homologous sequences to the siRNA (**Fig. 1.1**).

3. Methods to Generate siRNA in Cells

3.1. Long dsRNA

Because of the robust effect of siRNA to knockdown target gene expression, this technique is now used widely as a tool to screen gene function in many cell types, including stem cells. Strategies

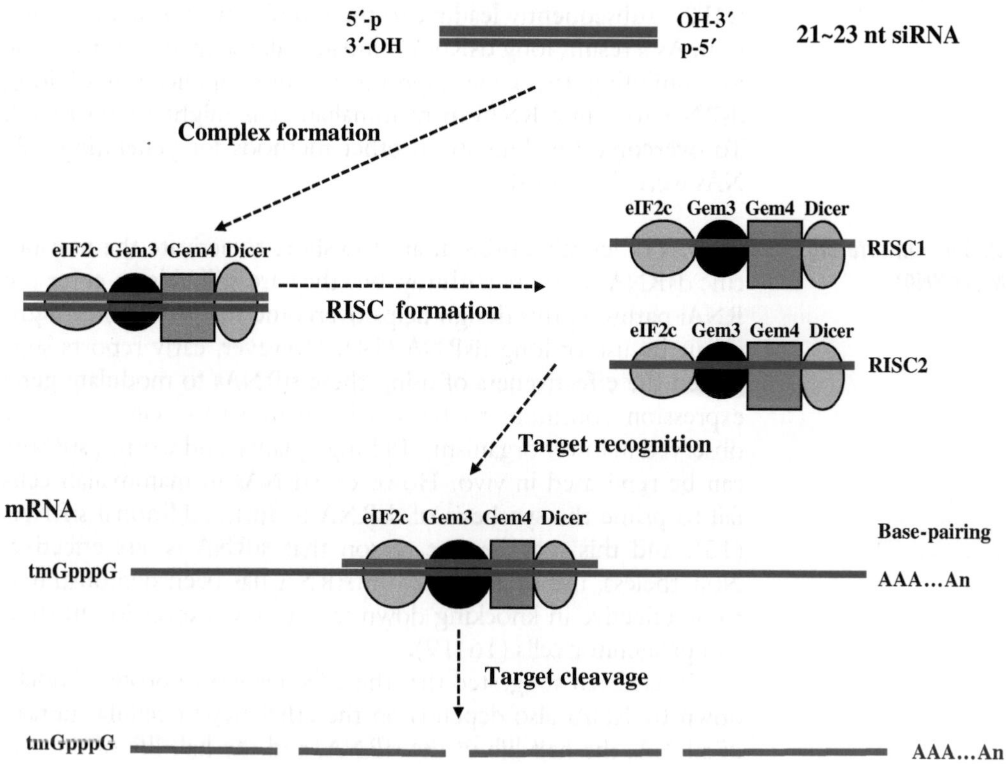

Fig. 1.1. Schematic representation of RISC formation and siRNA-mediated gene silencing. Small dsRNAs bind several proteins, such as eIF2c, GEM3, GEM4, and Dicer to form an siRNA–protein complex. Then the siRNA unwinds to a single strand; one of the unwound single strands of siRNA remains in this complex, named as RISC. The unwound single-strand RNA in RISC binds the target mRNA, which is consequently degraded by Dicer.

have been developed to knockdown individual or a group of genes using a combination of siRNAs. One approach to generate siRNAs is through the use of long dsRNAs. dsRNAs larger than 30 nts are normally referred to as long dsRNAs. It has been suggested that cellular uptake of long dsRNA induces RNAi in a diverse group of lower eukaryotic organisms, such as plants (8), Drosophila (9), and *C. elegans* (10). Long dsRNAs are cut by Dicer to form siRNAs when they are introduced into mammalian cells, and these siRNAs then should trigger RNAi. However, multiple attempts to induce RNAi, using long dsRNAs, in mammalian cell lines were met with limited success, due in part to the induction of the interferon response (11) or activation of a dsRNA-dependent protein kinase leading to nonspecific translational inhibition (12, 13). dsRNAs longer than 30 nts can trigger interferon I type responses and STAT-mediated dsRNA-dependent protein kinase (PKR) expression. This leads to the nonspecific degradation of mRNA and a general shutdown of host cell protein translation. Long dsRNAs may also activate PKR-mediated phosphorylation of the alpha-subunit of eukaryotic initiation factor-2 (eIF2), subsequently leading to the inhibition of protein translation. As a result, long dsRNA is intrinsically sequence-nonspecific for inhibiting the gene expression. Thus, application of long dsRNA to prime RNAi in mammalian cells might be restricted. To overcome this limitation, other methods for generating siRNAs were developed.

3.2. Small Interfering RNA (siRNA)

siRNAs of length 21–23 nt are too short to activate the nonspecific dsRNA response pathway, but they are still able to enter the RNAi pathway; this design helps overcome the problem brought about by use of long dsRNA (14). However, early reports suggested the effectiveness of using these siRNAs to modulate gene expression continues to be less in mammalian cells than that observed in lower organisms. In fungi, plants, and worms, siRNAs can be replicated in vivo. However, siRNAs in mammalian cells fail to prime the synthesis of dsRNA to form additional siRNAs (15), and this may be one reason that siRNA is less effective. Nonetheless, use of 21- to 23-nt siRNA has been demonstrated to be effective in knocking down target gene expression in stem and progenitor cells (16–19).

It has been suggested that the effectiveness of protein knockdown by RNAi also depends on the efficiency of cellular uptake of siRNA, the half-life of the siRNA, and the half-life of the targeted protein (20). If the protein expression level in the targeted cell is high and the half-life of the protein is long, it may prove to be difficult to knockdown expression of this particular protein in the 2 or 3 days that the siRNA generally persists. Therefore, additional modifications in methods to enhance siRNA applications have been made recently.

3.3. siRNA Pool

For the purpose of avoiding the phenomenon in which a single siRNA sequence may not efficiently knockdown a particular gene product, it has been suggested that at least three different siRNA sequences per target gene should be assessed for their ability to knockdown expression of the target gene (14). In addition, more robust knockdown of target gene expression can be reached by using a pool of siRNAs compared to a single siRNA (21). Knockdown of target gene expression for up to 5–10 days has been observed when using a pool of siRNAs in transfected cells. However, the cost also increases dramatically when siRNA pools, rather than single siRNAs, are applied.

3.4. DNA Vector–Mediated RNAi

Because of the transient nature of the gene-silencing effect invoked when using oligonucleotide siRNAs and the prohibitively high costs of chemical synthesis of these reagents, researchers have developed DNA plasmid vectors capable of expressing siRNAs intracellularly. Expression cassettes have been developed using the endogenous U6 snRNA or H1 RNA polymerase III promoters to drive expression of sequence-specific small hairpin RNAs (shRNAs) that stably regulate gene expression in mammalian cells via RNAi (22–24). These systems are based on the expression of siRNAs either as two separate strands or as a single shRNA (**Fig. 1.2**). It is suggested that these hairpin RNAs are processed by Dicer to active siRNAs in vivo (25). As an alternative approach, some groups have used the co-expression of sense and antisense RNA strands from independent expression cassettes or divergent cassettes (20, 26). The sense and antisense strands are believed to form a duplex in vivo, similar to the chemically synthesized siRNAs described previously (27). It has been suggested that the hairpin siRNA strategy appears to inhibit gene expression more efficiently than the duplex siRNAs expressed from two separate plasmids (24).

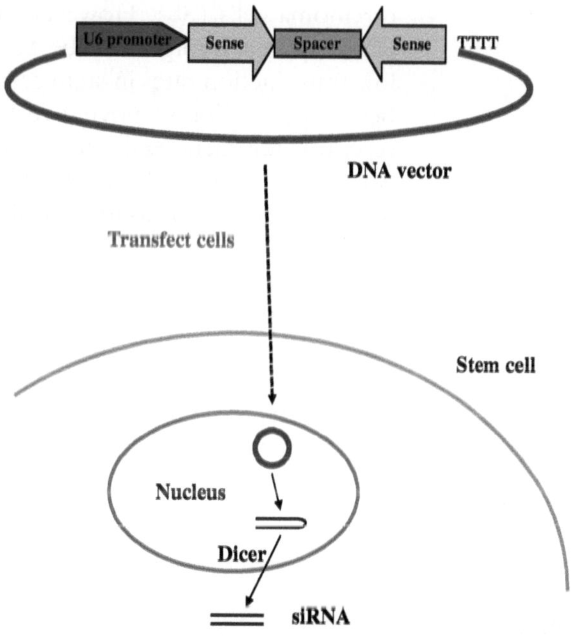

Fig. 1.2. DNA vector–mediated delivery and synthesis of siRNA. The U6 RNA Polymerase III promoter will direct transcription of a short hairpin (sh) RNA. The insert encoding the shRNA is cloned into the +1 position of the U6 promoter (−351 to + 1). The insert is composed of a 'sense' 19–23 nucleotide sequence that is complementary to the mRNA to which the siRNA will be targeted, a 6–9 nucleotide spacer, and an inverted repeat of the 'sense' sequence (thus, the two RNA sequences will anneal following nuclear synthesis to form the shRNA as depicted). Once inside the cell, the vector-derived shRNA will be constitutively expressed. As noted earlier, shRNAs (representing double-strand RNA) are recognized by Dicer inside the cells and the shRNAs are clipped to form siRNAs.

3.5. Virus Vector–Mediated RNAi

The recombinant viral vectors have been used to deliver shRNA in cultured mammalian cells (28–32). Lentiviral vectors have been developed to deliver shRNA into human 293 T cells (30), human ovary cancer cells (29), Hela cells (32), mouse hematopoietic stem cells (HSCs), and embryonic stem (ES) cells (28). shRNA introduced by lentiviral vectors could be useful for stable gene suppression in human cells (33). The advantage of using a lentiviral vector is that these vectors can be efficiently delivered and lead to a long-term stable expression of RNAi in both dividing and nondividing cells. Adenoviral vectors have also been reported recently to succeed in delivering siRNA to target cells. In an in vivo study, adenoviral vector-mediated beta-glucuronidase siRNA was effective in knocking down endogenous beta-glucuronidase expression in the mouse liver (34). The p53 gene protein product was also knocked down in 293 cells after adenoviral vector-mediated siRNA delivery (35). Adenoviral vector-mediated siRNA delivery has also been effective in knocking down target gene expression in MCF-7 cells (36), human umbilical vein endothelial cells (37), and human squamous cell carcinoma cells (38). However, this adenoviral vector-mediated RNAi approach might not be as applicable in stem cells because low transduction rates in stem cells, such as ES cell or HSCs, have been reported for adenoviral vectors. It appears that the primary receptors for adenoviral viruses are poorly expressed in stem cells. Adenoviral-associated vectors (AAVs) have been successfully used to deliver RNAi to nonstem cells. Ma et al. (39) used AAVs to deliver siRNA into BHK-21 cells and diminished the expression of a luciferase reporter gene by 70%. Several investigators have reported effective delivery of shRNA using recombinant AAV vectors with appropriate gene expression inhibition in nonstem cells (40, 41). However, as AAVs also have low transduction rates in many stem cell populations, further investigations will be required to determine if this is an effective strategy for RNAi delivery.

4. RNAi Applications to Study Stem Cell Function

4.1. RNAi Studies in ES Cells

Stem cells are distinguished from other cell types by two important characteristics. First, they are unspecialized cells capable of renewing themselves through cell division, sometimes after long periods of inactivity. Second, under certain physiologic or experimental conditions, they can be induced to become tissue- or organ-specific cells by differentiation with special functions. ES cells are pluripotent stem cells derived from the inner cell mass of the 3.5-day-old mouse blastocyst (42, 43). These cells are an attractive model to study the molecular regulation of cell lineage commitment and cellular differentiation because ES cells can give rise to cells derived from all three primary germ layers: endoderm, mesoderm, and ectoderm. Yang and his colleagues (44) demonstrated the ability to diminish expression of a target gene in undifferentiated ES cells by in situ production of long dsRNA from a transient transfection of a plasmid harboring a 547-bp inverted repeat or direct transfection of a 740-bp dsRNA made by in vitro transcription. However, these long dsRNAs could only mediate RNAi in undifferentiated ES cells, but not differentiated ES cells. This difference in RNAi efficacy between undifferentiated and differentiated cells when using dsRNA remains unexplained. As an alternative approach, we transfected well-differentiated ES cells with siRNA and found that these oligonucleotides were effective in diminishing the expression of such genes as *PU.1* and *c-EBPα* (17). More recently, we have successfully blocked the expression of Shp-2, a protein tyrosine phosphatase, in differentiated murine ES cells resulting in the reduction of hemangioblast development (18). We also investigated suppression of Ape1, a redox protein in ES cells that affects their hematopoietic differentiation (19). *Oct-4* is an important gene in sustaining pluripotency in ES cells. Velkey and O'Shea (45) examined whether suppression of *Oct-4* expression via RNAi would alter ES cell lineage commitment decision. In their study, ES cells were transfected with plasmids containing an independently expressed reporter gene and an RNA polymerase type III promoter to constitutively express small stem-loop RNA transcripts corresponding to Oct-4 mRNA. Cells transfected with Oct-4 shRNA demonstrated reduced levels of Oct-4 mRNA and exhibited characteristics of trophectodermal differentiation. More recently, Oct-4 siRNA delivered by transfection was effective in both human and mouse ES cells in diminishing *Oct-4* expression (46).

4.2. RNAi Studies in Hematopoietic Stem Cells

HSCs represent a self-renewing population of rare cells that give rise to all differentiated hematopoietic elements (47). Use of RNAi in primary cells such as HSCs would facilitate rapid gene discovery in a postgenome era (16). Several RNAi approaches can be considered in strategies to discover gene functions in HSCs. For example, when CD45 siRNA was delivered into murine bone marrow Sca-l$^+$ hematopoietic progenitors via electroporation to inhibit CD45 expression, 3-fold more hematopoietic colonies were detected in a progenitor assay (16). Akkina et al. (48) introduced the HIV Rev and Tat specific gene siRNA into CD34$^+$ hematopoietic progenitor cells via lentiviral vectors. These siRNA transduced CD34$^+$stem cells could differentiate normally into HIV-1 resistant macrophages in vitro and T cells in vivo in SCID-hu mice following transplantation. These data have demonstrated the utility of siRNAs delivered into HSCs via lentiviral vectors for future in vivo applications (49).

Recently, it has also been suggested that siRNA exhibited the ability to knockdown target gene expression in human HSCs. For instance, Scherr et al. (50) demonstrated that stable RNAi can be induced in human HSCs by lentiviral gene transfer of shRNAs that then mediate long-term gene silencing of an endogenous hematopoietic-specific gene. They also transfected normal human primary $CD34^{+}$hematopoietic cells and $CD34^{+}$ chronic myeloid leukemia cells with bcr-abl siRNA and found these cells were also sensitive to siRNA targeting (51). This study revealed that siRNA is a potential tool to specifically and efficiently reduce the expression of an oncogenic fusion gene in hematopoietic cells. siRNA might offer a new tool to study oncogene function in human HSCs. Furthermore, this approach might be a potential reagent for treating hematopoietic malignancies in the future via altering expression levels of a specific malignant gene.

4.3. RNAi in Cancer Stem Cells: Stem Cell Disease and RNAi Therapy

Cancer stem cells (CSCs) are cancer cells (found within tumors or hematological cancers) that possess characteristics associated with normal stem cells, specifically the ability to give rise to all cell types found in a particular cancer sample. Chronic myeloid leukemia (CML) can be treated with a tyrosine kinase inhibitor, such as imatinib. However, early relapses are still problematic. It has been suggested that imatinib may not effectively eradicate leukemic stem/progenitor cells. Abelson helper integration site 1 (*Ahi-1*)/*AHI-1* is an oncogene that is highly deregulated in CML stem/progenitor cells where levels of BCR-ABL transcripts are also elevated. Overexpression of *Ahi-1*/*AHI-1* in murine and human hematopoietic cells confers growth advantages in vitro and induces leukemia in vivo, enhancing effects of BCR-ABL. Conversely, RNAi-mediated suppression of *AHI-1* in BCR-ABL-transduced lin(−)CD34(+) human cord blood cells and primary CML stem/progenitor cells reduced their growth autonomy in vitro. Moreover, co-expression of *Ahi-1* in BCR-ABL-inducible cells reverses growth deficiencies exhibited by BCR-ABL down-regulation and is associated with sustained phosphorylation of BCR-ABL and enhanced activation of JAK2-STAT5 (52). These studies implicate *AHI-1* as a potential therapeutic target downstream of BCR-ABL in CML, and *AHI-1* RNAi-mediated suppression of *AHI-1* gene expression may be a potential in CML therapy.

5. Conclusion and Perspectives

Stem cells exhibit strong potential in regenerative medicine and cancer biology. ES cells represent a powerful reagent to understand the molecular pathways that result in the cells of the three germ layers (endoderm, mesoderm, and ectoderm) giving rise to a host of differentiated cell types. Techniques continue to be developed that permit directed cellular differentiation of ES cells into specific cell lineages. Therefore, the ES cell system provides a possible strategy for the high-throughput functional screening of genes that are required for specific differentiation of ES cells into a specific cell lineage.

One of the most widely utilized approaches to the discovery of murine gene function is via the use of targeted gene inactivation. Murine ES cells have proven to be an important reagent as the genome of these cells can be manipulated in vitro leading to the inactivation of at least one allele of a gene of interest. In vitro selection of the affected ES cells can be performed via several methods resulting in a population of mutant cells. These ES cells can be injected into blastocysts to create founder chimeric mice that carry the mutant allele in the germ line. Subsequently, these founders can be bred and eventually both heterozygous and homozygous mutant animals derived. This strategy has led to the discovery of the function of thousands of murine genes. Nonetheless, this represents a small proportion of the total number of assumed murine genes based upon sequence data and the methods are laborious and expensive.

It is obvious that the use of RNAi to knockdown gene expression in ES cells may provide a novel screening approach to understand gene function that is amenable for large-scale, high-throughput methods. RNAi approaches do not require that the entire gene sequence be known at the time of study. Thus, RNAi targeted to specific expressed sequence tags (cDNA encoding for portions of unknown genes) or entire sequenced genes of unknown function could be utilized to discover the consequences of diminished expression of the gene products and the roles these molecules play in cellular differentiation.

Recently, a collection of several hundred genes have been identified in hematopoietic stem cells, neural stem cells, and ES cells that appear to be shared. Whether these genes are specifically required for maintenance of 'stemness' is being determined. RNAi may provide an approach to the knockdown of expression of the protein products of these genes in these stem cell populations and new insights may emerge in our understanding of the molecular regulation of 'stemness' in cancer stem cells – consequently, bringing a new perspective in cancer therapy through RNAi-mediated target gene knockdown in cancer stem cells.

Acknowledgements

This work was supported by NIH grant CA84405 and the Spastic Paralysis Foundation of the Illinois-Eastern Iowa Division

Kiwanis International. We are grateful to Dr. Janet Rowley of the University of Chicago and Drs Mervin Yoder and Mark Kelley of Indiana University for collaboration and support in these works.

References

1. Fire, A., Xu, S., Montgomery, M. K., Kostas, S. A., Driver, S. E., and Mello, C. C. (1998) Potent and specific genetic interference by double-stranded RNA in *Caenorhabditis elegans*. *Nature* **391**, 806–811.
2. Guo, S., and Kemphues, K. J. (1995) *par-1*, a gene required for establishing polarity in *C. elegans* embryos, encodes a putative Ser/Thr kinase that is asymmetrically distributed. *Cell* **81**, 611–620.
3. Shi, Y. (2003) Mammalian RNAi for the masses. *Trends Genet* **19**, 9–12.
4. Zamore, P. D., Tuschl, T., Sharp, P. A., and Bartel, D. P. (2000) RNAi: Double-stranded RNA directs the ATP-dependent cleavage of mRNA at 21–23 nucleotide intervals. *Cell* **101**, 25–33.
5. Zamore, P. D. (2002) Ancient pathways programmed by small RNAs. *Science* **296**, 1265–1269.
6. Nykanen, A., Haley, B., and Zamore, P. D. (2001) ATP requirements and small interfering RNA structure in the RNA interference pathway. *Cell* **107**, 309–321.
7. Tuschl, T. (2001) RNA interference and small interfering RNAs. *Chembiochem* **2**, 239–245.
8. Hamilton, A. J., and Baulcombe, D. C. (1999) A species of small antisense RNA in posttranscriptional gene silencing in plants. *Science* **286**, 950–952.
9. Kennerdell, J. R., and Carthew, R. W. (2000) Heritable gene silencing in Drosophila using double-stranded RNA. *Nat Biotechnol* **18**, 896–898.
10. Tabara, H., Sarkissian, M., Kelly, W. G., Fleenor, J., Grishok, A., Timmons, L., Fire, A., and Mello, C. C. (1999) The *rde-1* gene, RNA interference, and transposon silencing in *C. elegans*. *Cell* **99**, 123–132.
11. Ui-Tei, K., Zenno, S., Miyata, Y., and Saigo, K. (2000) Sensitive assay of RNA interference in Drosophila and Chinese hamster cultured cells using firefly luciferase gene as target. *FEBS Lett* **479**, 79–82.
12. Gil, J., and Esteban, M. (2000) Induction of apoptosis by the dsRNA-dependent protein kinase (PKR): Mechanism of action. *Apoptosis* **5**, 107–114.
13. Kumar, M., and Carmichael, G. G. (1998) Antisense RNA: Function and fate of duplex RNA in cells of higher eukaryotes. *Microbiol Mol Biol Rev* **62**, 1415–1434.
14. Caplen, N. J., and Mousses, S. (2003) Short interfering RNA (siRNA)-mediated RNA interference (RNAi) in human cells. *Ann N Y Acad Sci* **1002**, 56–62.
15. Dykxhoorn, D. M., Novina, C. D., and Sharp, P. A. (2003) Killing the messenger: Short RNAs that silence gene expression. *Nat Rev Mol Cell Biol* **4**, 457–467.
16. Oliveira, D. M., and Goodell, M. A. (2003) Transient RNA interference in hematopoietic progenitors with functional consequences. *Genesis* **36**, 203–208.
17. Zou, G. M., Wu, W., Chen, J., and Rowley, J. D. (2003) Duplexes of 21-nucleotide RNAs mediate RNA interference in differentiated mouse ES cells. *Biol Cell* **95**, 365–371.
18. Zou, G. M., Chan, R. J., Shelley, W. C., and Yoder, M. C. (2006) Reduction of Shp-2 expression by small interfering RNA reduces murine embryonic stem cell-derived in vitro hematopoietic differentiation. *Stem Cells* **24**, 587–594.
19. Zou, G. M., Luo, M. H., Reed, A., Kelley, M. R., and Yoder, M. C. (2007) Ape1 regulates hematopoietic differentiation of embryonic stem cells through its redox functional domain. *Blood* **109**, 1917–1922.
20. Wang, J., Tekle, E., Oubrahim, H., Mieyal, J. J., Stadtman, E. R., and Chock, P. B. (2003) Stable and controllable RNA interference: Investigating the physiological function of glutathionylated actin. *Proc Natl Acad Sci USA* **100**, 5103–5106.
21. Michaelson, J. S., and Leder, P. (2003) RNAi reveals anti-apoptotic and transcriptionally repressive activities of DAXX. *J Cell Sci* **116**, 345–352.
22. Miyagishi, M., and Taira, K. (2002) U6 promoter-driven siRNAs with four uridine 3′ overhangs efficiently suppress targeted gene expression in mammalian cells. *Nat Biotechnol* **20**, 497–500.
23. Paul, C. P., Good, P. D., Winer, I., and Engelke, D. R. (2002) Effective expression of small interfering RNA in human cells. *Nat Biotechnol* **20**, 505–508.
24. Yu, J. Y., DeRuiter, S. L., and Turner, D. L. (2002) RNA interference by expression of

short-interfering RNAs and hairpin RNAs in mammalian cells. *Proc Natl Acad Sci USA* **99**, 6047–6052.

25. Paddison, P. J., Caudy, A. A., and Hannon, G. J. (2002) Stable suppression of gene expression by RNAi in mammalian cells. *Proc Natl Acad Sci USA* **99**, 1443–1448.
26. Lee, N. S., Dohjima, T., Bauer, G., Li, H., Li, M. J., Ehsani, A., Salvaterra, P., and Rossi, J. (2002) Expression of small interfering RNAs targeted against HIV-1 rev transcripts in human cells. *Nat Biotechnol* **20**, 500–505.
27. Elbashir, S. M., Lendeckel, W., and Tuschl, T. (2001) RNA interference is mediated by 21- and 22-nucleotide RNAs. *Genes Dev* **15**, 188–200.
28. Rubinson, D. A., Dillon, C. P., Kwiatkowski, A. V., Sievers, C., Yang, L., Kopinja, J., Rooney, D. L., Zhang, M., Ihrig, M. M., McManus, M. T., Gertler, F. B., Scott, M. L., and Van Parijs, L. (2003) A lentivirus-based system to functionally silence genes in primary mammalian cells, stem cells and transgenic mice by RNA interference. *Nat Genet* **33**, 401–406.
29. Yang, G., Thompson, J. A., Fang, B., and Liu, J. (2003) Silencing of H-ras gene expression by retrovirus-mediated siRNA decreases transformation efficiency and tumor growth in a model of human ovarian cancer. *Oncogene* **22**, 5694–5701.
30. An, D. S., Xie, Y., Mao, S. H., Morizono, K., Kung, S. K., and Chen, I. S. (2003) Efficient lentiviral vectors for short hairpin RNA delivery into human cells. *Hum Gene Ther* **14**, 1207–1212.
31. Brummelkamp, T. R., Bernards, R., and Agami, R. (2002) A system for stable expression of short interfering RNAs in mammalian cells. *Science* **296**, 550–553.
32. Devroe, E., and Silver, P. A. (2002) Retrovirus-delivered siRNA. *BMC Biotechnol* **2**, 15.
33. Nishitsuji, H., Ikeda, T., Miyoshi, H., Ohashi, T., Kannagi, M., and Masuda, T. (2004) Expression of small hairpin RNA by lentivirus-based vector confers efficient and stable gene-suppression of HIV-1 on human cells including primary non-dividing cells. *Microbes Infect* **6**, 76–85.
34. Xia, H., Mao, Q., Paulson, H. L., and Davidson, B. L. (2002) siRNA-mediated gene silencing in vitro and in vivo. *Nat Biotechnol* **20**, 1006–1010.
35. Zhao, L. J., Jian, H., and Zhu, H. (2003) Specific gene inhibition by adenovirus-mediated expression of small interfering RNA. *Gene* **316**, 137–141.
36. Shen, C., Buck, A. K., Liu, X., Winkler, M., and Reske, S. N. (2003) Gene silencing by adenovirus-delivered siRNA. *FEBS Lett* **539**, 111–114.
37. Arts, G. J., Langemeijer, E., Tissingh, R., Ma, L., Pavliska, H., Dokic, K., Dooijes, R., Mesic, E., Clasen, R., Michiels, F., van der Schueren, J., Lambrecht, M., Herman, S., Brys, R., Thys, K., Hoffmann, M., Tomme, P., and van Es, H. (2003) Adenoviral vectors expressing siRNAs for discovery and validation of gene function. *Genome Res* **13**, 2325–2332.
38. Shin, J. J., Katayama, T., Michaud, W. A., and Rocco, J. W. (2004) Short hairpin RNA system to inhibit human p16 in squamous cell carcinoma. *Arch Otolaryngol Head Neck Surg* **130**, 68–73.
39. Ma, X., Lu, X., Peng, J., Tan, S., Yuan, Z., Yin, F., Wang, H., Wu, X., and Hou, Y. (2002) [Inhibition of Luciferase expression in mammalian cells by AAV vector plasmid mediated Luciferase shRNA]. *Zhonghua Shi Yan He Lin Chuang Bing Du Xue Za Zhi* **16**, 253–255.
40. Boden, D., Pusch, O., Silbermann, R., Lee, F., Tucker, L., and Ramratnam, B. (2004) Enhanced gene silencing of HIV-1 specific siRNA using microRNA designed hairpins. *Nucleic Acids Res* **32**, 1154–1158.
41. Tomar, R. S., Matta, H., and Chaudhary, P. M. (2003) Use of adeno-associated viral vector for delivery of small interfering RNA. *Oncogene* **22**, 5712–5715.
42. Evans, M. J., and Kaufman, M. H. (1981) Establishment in culture of pluripotential cells from mouse embryos. *Nature* **292**, 154–156.
43. Zou, G. M., Reznikoff-Etievant, M. F., Hirsch, F., and Milliez, J. (2000) IFN-gamma induces apoptosis in mouse embryonic stem cells, a putative mechanism of its embryotoxicity. *Dev Growth Differ* **42**, 257–264.
44. Yang, S., Tutton, S., Pierce, E., and Yoon, K. (2001) Specific double-stranded RNA interference in undifferentiated mouse embryonic stem cells. *Mol Cell Biol* **21**, 7807–7816.
45. Velkey, J. M., and O'Shea, K. S. (2003) Oct-4 RNA interference induces trophectoderm differentiation in mouse embryonic stem cells. *Genesis* **37**, 18–24.
46. Hay, D. C., Sutherland, L., Clark, J., and Burdon, T. (2004) Oct-4 knockdown induces similar patterns of endoderm and trophoblast differentiation markers in human and mouse embryonic stem cells. *Stem Cells* **22**, 225–235.

47. Yoder, M. C. (2004) Blood cell progenitors: Insights into the properties of stem cells. *Anat Rec A Discov Mol Cell Evol Biol* **276**, 66–74.
48. Akkina, R., Banerjea, A., Bai, J., Anderson, J., Li, M. J., and Rossi, J. (2003) siRNAs, ribozymes and RNA decoys in modeling stem cell-based gene therapy for HIV/AIDS. *Anticancer Res* **23**, 1997–2005.
49. Banerjea, A., Li, M. J., Bauer, G., Remling, L., Lee, N. S., Rossi, J., and Akkina, R. (2003) Inhibition of HIV-1 by lentiviral vector-transduced siRNAs in T lymphocytes differentiated in SCID-hu mice and $CD34^+$ progenitor cell-derived macrophages. *Mol Ther* **8**, 62–71.
50. Scherr, M., Battmer, K., Dallmann, I., Ganser, A., and Eder, M. (2003) Inhibition of GM-CSF receptor function by stable RNA interference in a NOD/SCID mouse hematopoietic stem cell transplantation model. *Oligonucleotides* **13**, 353–363.
51. Scherr, M., Battmer, K., Winkler, T., Heidenreich, O., Ganser, A., and Eder, M. (2003) Specific inhibition of bcr-abl gene expression by small interfering RNA. *Blood* **101**, 1566–1569.
52. Zhou, L. L., Zhao, Y., Ringrose, A., DeGeer, D., Kennah, E., Lin, A. E., Sheng, G., Li, X. J., Turhan, A., and Jiang, X. (2008) AHI-1 interacts with BCR-ABL and modulates BCR-ABL transforming activity and imatinib response of CML stem/progenitor cells. *J Exp Med* **205**, 2657–2671.

Chapter 2

A Brief Introduction to RNAi and MicroRNAs in Stem Cells

Alexander K. Murashov

Abstract

Recently, RNAi, including microRNAs (miRNAs), has become an important tool to investigate the regulatory mechanism of stem cell maintenance and differentiation. In this short chapter, we will give a brief overview of the discovery history, functions, and mechanisms of RNAi and miRNAs. We will also discuss RNAi as a tool to study stem cell function and the potential future practical applications.

Key words: RNAi, siRNA, miRNA, gene regulation, stem cell.

1. A Short History of RNAi

RNA interference (RNAi) was a term to describe the inhibition of gene expression by double-stranded RNAs (dsRNAs). RNAi was originally discovered in plants as co-suppression (1) and fungi as quelling (2), which lead to simultaneous silencing of an experimentally introduced gene and a homologous endogenous gene (3). This mechanism was first specifically characterized in the worm species *Caenorhabditis elegans*, when injection of double-stranded RNA (dsRNA) several hundred nucleotides long was found to inhibit the expression of corresponding gene products (4). Subsequent experiments showed that long dsRNA could also induce a gene-specific inhibition of expression in a number of invertebrates including *Drosophila* (5, 6). Further studies showed that RNAi is an evolutionarily conserved pathway of post-transcriptional gene silencing, which exists among virtually all eukaryotes, including mammals (7). While initial introduction of long dsRNAs into mammalian cells was unsuccessful, later

B. Zhang, E.J. Stellwag (eds.), *RNAi and MicroRNA-Mediated Gene Regulation in Stem Cells*, Methods in Molecular Biology 650, DOI 10.1007/978-1-60761-769-3_2, © Springer Science+Business Media, LLC 2010

experiments showed that short dsRNAs 21–23 nucleotides long, termed siRNAs, could suppress mammalian gene expression in a highly specific manner, paving the way for practical gene silencing in mammalian cells (8). In recent years, RNAi has developed as an effective tool to specifically knockdown gene expression in a wide variety of target cells. Delivering siRNA to target cells in vivo has been problematic, and therefore, most of the current information on RNAi is derived from in vitro studies and epigenetic approaches. Current research efforts in this area are focused on developing approaches to deliver siRNA in vivo and to characterize RNAi in specific mammalian cell types. Several studies have demonstrated siRNA-mediated inhibition of gene expression in stem cells (9).

2. RNAi Mechanism

RNAi is a natural mechanism that acts to selectively suppress gene expression (3, 10, 11). The RNAi machinery appears to have emerged early in evolution to protect the eukaryotic genome from endogenous transposable elements and from viral infections (12). Recent observations demonstrated that in addition to its protective action, RNAi plays an important role during cell growth and differentiation (13) and in early development (14, 15). RNAi is usually initiated by short interfering double-strand RNAs (siRNAs) or native small single-strand microRNAs (miRNAs). Both siRNA and miRNA enter RNA-induced silencing complex (RISC) which is responsible for the sequence-specific degradation of the target RNAs (16).

RISC contains the molecular machinery to recognize and destroy target mRNAs, thereby preventing protein synthesis. Thus, RISC is frequently referred to as the effector mechanism in the RNAi pathway (17, 18). Although the components of the mammalian RISC complex remain under investigation, several subunits have been conclusively identified to date: Argonaute2 (Ago2) nuclease (18, 19), fragile X mental retardation protein (FMRP) (20), a Tudor-SN nuclease (21), Gemin3, Gemin4 (22–24), TRBP (the human immunodeficiency virus transactivating response RNA-binding protein) (25), dsRNA-binding protein PACT (26), and a general translation repressor protein RCK (also known as p54) (27). Here we briefly describe the most well-studied components. The Argonaute2 (Ago2) nuclease, which destroys mRNA by cleaving the bonds between adjacent nucleotides located directly across from the center of the guide siRNA, is encoded by a member of the Argonaute gene family (28). The number of Argonaute proteins differs among species. For example, humans have eight Argonaute proteins (29) while mice have eight and *Drosophila* have five (30). Thus far, Ago2 is the sole Argonaute protein that has been proven to be a *bona fide* component of RISC (31). It has further been shown that Ago2 plays a vital role in RNAi, as evidenced by a pronounced reduction in RNAi after knockdown of Ago2 protein (18).

The fragile X gene family is located on the X chromosome and encodes three different proteins: fragile X mental retardation protein (FMRP, also called FXMRP and FMR1), FXR1, and FXR2 (32, 33). Absence or deficit of FMRP resulting from gene mutations, is the fundamental defect in fragile X syndrome (FXS). Although much is known about the function of FMRP, the functions of FXR1 and FXR2 are unclear. It is suspected, however, that all three fragile X proteins bind with ribosomes (predominately the 60S ribosomal subunits) and play a potential role in translation or mRNA stability (20). Interestingly, although it appears that FMRP is a component of the RISC complex, suppression of FMRP does not disrupt RNAi, suggesting that FMRP has some unknown accessory function in the RISC complex (20, 33). The extent to which FXR1 and FXR2 are associated with RISC is also unclear; however, it is likely that they may have a role similar to FMRP.

Researchers speculate that another component of RISC may be a nuclease (34–36). Recent observation suggests that the additional nuclease component of RISC may be tudor staphylococcal nuclease (Tudor-SN, also known as the mammalian homologue p100), which is a protein encoded by the micrococcal nuclease gene family (21). The study demonstrated that p100 forms a complex with the other identified protein components of RISC (i.e., Ago2, FMRP). Moreover, p100 also showed nuclease activity. However, the role of p100/Tudor-SN in cleavage of mRNA has not yet been established.

Recent observations revealed that mammalian RISC also contains the micro-ribonucleoprotein complex (miRNP) proteins eIF2C2 (the human homologue of Ago2), Gemin3, and Gemin4 (22–24). Gemin3 is a DEAD-box RNA helicase that binds to the survival of motor neurons (SMN) protein. Gemin4 is also a component of the SMN complex (22). Interestingly, reduction in SMN protein results in spinal muscular atrophy (SMA), a common neurodegenerative disease. The SMN complex has critical functions in the assembly/restructuring of diverse ribonucleoprotein (RNP) complexes in the nervous system. These findings suggest that miRNP proteins, including eIF2C2/Ago2, Gemin3, and Gemin4, may play important roles in target mRNA recognition and translational repression in neurons (24). While recent research demonstrated that Gemin3 and Gemin4 form a complex with eIF2C2/Ago2 in neurons, whether or not the complex includes the other known components of RISC remains to be verified. Additional research is needed to determine the exact association and functional role of Gemin3 and Gemin4 in RISC, and whether or not any interplay exists between the components of RISC and neuronal multi-protein complexes implicated in SMA and/or FXS.

3. MiRNAs and Stem Cell

MiRNAs are a newly identified class of endogenous small RNAs (37–40), with varied length from 16 to 29 nt and a majority of them from 21–23 nt (41). MiRNAs biogenesis consists of several key steps including processing by Drosha, DGCR8/Pasha, Exportin5, Dicer, RISC proteins, and P-bodies (16). In mammals, miRNA biogenesis starts with primary miRNA (pri-miRNA) which is processed by Drosha and DGCR8 (DiGeorge syndrome critical region gene-8)/Pasha into ~70 nt precursor-miRNA (pre-miRNA) (42). Pre-miRNA then exits the nucleus via the action of a nuclear transport receptor complex, Exportin-5 (Exp5)-RanGTP (43). Dicer, a cytoplasmic RNase III nuclease, then cleaves pre-miRNA to generate ~22 nt miRNA:miRNA duplex (44). One strand (miRNA) of the duplex is then loaded onto RISC for targeting gene expression whereas miRNA, in most cases, is degraded by an unknown mechanism (25, 45). RISC associated with mRNA is either stored in discrete cytoplasmic domains known as processing bodies (P-bodies) or enter the mRNA-decay pathway for destruction (46).

Although the first miRNA (let-7) was discovered 25 years ago in *C. elegans* (47, 48), it was not recognized until the early 2000s (49–51). Since then, miRNAs have attracted considerable attention from scientists and currently the miRNA-related field has become one of the hottest research topics in biomedicine. More and more experimental studies demonstrated that miRNAs play important and diverse roles in almost all biological and metabolic processes, including early development (52), cell proliferation (53), cell death (54), fat metabolism (55), signal transduction (56, 57), and diseases (58, 59). Disregulation of miRNAs expression has been implicated in developmental defects (53, 60), cancers (61, 62), and nervous system diseases (63). A number of miRNAs were found in the vertebrate nervous system (64, 65). Moreover, a recent observation has revealed an important role for the miRNAs in zebrafish brain development (66), as well as in later stages of mammalian neuronal maturation and synapse formation (67–69). Recent observation by the Greenberg group has demonstrated that brain-specific miRNA miR-134 can regulate local dendritic translation of LIM-domain kinase 1 (Limk1) and thus play a role in actin filament dynamics during dendritic spine development.

More interestingly, recent studies indicate that the expression profiles of miRNAs in stem cells are different from other tissues. This phenomenon plus other emerging evidence suggests that miRNAs may play an essential role in stem cell self-renewal and differentiation (70–90). MiRNAs appear to regulate the expression of a significant percentage of all genes in a wide array of mammalian cell types including stem cells (91). Embryonic stem (ES) cells are pluripotent stem cells derived from the inner cell mass of the 3.5-day-old mouse blastocyst. ES cells can give rise to cells derived from all three primary germ layers, endoderm, mesoderm, and ectoderm. ES cells are a popular model to study the molecular mechanisms of cellular differentiation. A subset of miRNAs is preferentially expressed in ES cells or embryonic tissue (92). Dicer-deficient mice fail to develop (93) and ES cells deficient in miRNA-processing enzymes show defects in differentiation (94). Specific miRNAs have been shown to participate in mammalian cellular differentiation and embryonic development (95). However, how transcription factors and miRNAs function together in the regulatory circuitry that controls early development remains to be answered. Recent observation suggests that miRNAs, which are activated in ES cells by Oct4/Sox2/Nanog/Tcf3, serve to modulate the direct effects of these transcription factors, participating in regulation to tune levels of key genes and modifying the gene expression program to help poise ES cells for efficient differentiation (91). Loss of miRNA pathway components negatively affects differentiation of ES cells, but the underlying molecular mechanisms remain poorly defined. A recent report characterized changes in mouse ES cells lacking Dicer. Transcriptome analysis of *Dicer*$^{-/-}$ cells indicated that the ES-specific miR-290 cluster had an important regulatory function in undifferentiated ES cells (96). Consistently, many of the defects in Dicer-deficient cells could be reversed by transfection with miR-290 family miRNAs. *Oct4* is a transcription factor which has been characterized as a key regulator of ES cell pluripotency. Observations showed that the expression level of *Oct4* was important in early lineage commitment of ES cells. *Oct4*-deficient embryos fail to form an inner cell mass, but remaining cells commit to the trophoblast lineage (97). Cells transfected with *Oct4* shRNA demonstrated reduced levels of *Oct4* transcription and exhibited characteristics of trophectodermal differentiation. In addition, transfection with *Oct4* siRNAs was effective in both mouse and human ES cells in silencing *Oct4* (98).

Interestingly, *Oct4* silencing in differentiating *Dicer*$^{-/-}$ ES cells was also accompanied by accumulation of repressive histone marks but not by DNA methylation, which prevents the stable repression of *Oct4* (96).

4. RNAi Methods to Study Stem Cell Function

The methods to study stem cell function include long dsRNAs, siRNAs, siRNA pools, DNA-vector-mediated RNAi, and virus-vector-mediated RNAi. Long dsRNAs are cut by Dicer to form siRNA when they are introduced into mammalian cells, and these siRNAs then trigger RNAi. However, multiple attempts to induce RNAi using long dsRNA in mammalian cell lines were met with limited success, due in part to the induction of interferon (99) or activation of a dsRNA-dependent protein kinase leading to nonspecific translational inhibition (100). siRNAs are too small to induce the nonspecific dsRNA response pathway, but they are still able to enter the RNAi pathway. However, the effectiveness of using these siRNAs to modulate gene expression continues to be less in mammalian cells than that observed in lower organisms. The effectiveness of knockdown by siRNAs also depends on the efficiency of uptake of siRNA and the half-life. Therefore, additional modifications to siRNA to enhance its stability and uptake have been made recently. Some investigators have suggested that a pool of at least three different siRNA sequences per target gene should be assessed for their ability to knock down expression of the target gene (101). Several reports demonstrated better knockdown of target genes using a pool of siRNAs compared with a single siRNA (102). Different recombinant viral vectors have been used to deliver shRNA to mammalian cells (103).

5. Conclusion and Perspectives

The differentiation of the ES lineages is a critical event when cell fate decisions are made and loss of pluripotency occurs. Application of RNAi in ES cells will provide important tools for the study of cell and tissue differentiation. RNAi, and particularly recently identified miRNAs, may provide new approaches to facilitate the knockdown of expression of the genes in different ES cell populations and provide new insights in our understanding of 'stemness.' It is a process about which much remains to be learned and many mechanisms remain to be elucidated. The power of

RNAi-based approaches can be applied for understanding ES cell gene function and serve to establish foundations in ES cell-based therapies.

References

1. Napoli, C., Lemieux, C., and Jorgensen, R. (1990) Introduction of a chimeric chalcone synthase gene into petunia results in reversible co-suppression of homologous genes in trans. *Plant Cell* **2**, 279–289.
2. Romano, N., and Macino, G. (1992) Quelling: Transient inactivation of gene expression in *Neurospora crassa* by transformation with homologous sequences. *Mol Microbiol* **6**, 3343–3353.
3. Hannon, G. J. (2002) RNA interference. *Nature* **418**, 244–251.
4. Fire, A., Xu, S., Montgomery, M. K., Kostas, S. A., Driver, S. E., and Mello, C. C. (1998) Potent and specific genetic interference by double-stranded RNA in *Caenorhabditis elegans. Nature* **391**, 806–811.
5. Boutla, A., Delidakis, C., Livadaras, I., Tsagris, M., and Tabler, M. (2001) Short 5′-phosphorylated double-stranded RNAs induce RNA interference in Drosophila. *Curr Biol* **11**, 1776–1780.
6. Kennerdell, J. R., and Carthew, R. W. (1998) Use of dsRNA-mediated genetic interference to demonstrate that frizzled and frizzled 2 act in the wingless pathway. *Cell* **95**, 1017–1026.
7. Duxbury, M. S., and Whang, E. E. (2004) RNA interference: A practical approach. *J Surg Res* **117**, 339–344.
8. Elbashir, S. M., Harborth, J., Lendeckel, W., Yalcin, A., Weber, K., and Tuschl, T. (2001) Duplexes of 21-nucleotide RNAs mediate RNA interference in cultured mammalian cells. *Nature* **411**, 494–498.
9. Chen, S., Choo, A., Wang, N. D., Too, H. P., and Oh, S. K. (2007) Establishing efficient siRNA knockdown in mouse embryonic stem cells. *Biotechnol Lett* **29**, 261–265.
10. Ambros, V. (2003) MicroRNA pathways in flies and worms: Growth, death, fat, stress, and timing. *Cell* **113**, 673–676.
11. Cerutti, H. (2003) RNA interference: Traveling in the cell and gaining functions?, *Trends Genet* **19**, 39–46.
12. Shi, Y. (2003) Mammalian RNAi for the masses. *Trends Genet* **19**, 9–12.
13. McManus, M. T., and Sharp, P. A. (2002) Gene silencing in mammals by small interfering RNAs. *Nat Rev Genet* **3**, 737–747.
14. Carrington, J. C., and Ambros, V. (2003) Role of microRNAs in plant and animal development. *Science* **301**, 336–338.
15. Wienholds, E., Koudijs, M. J., van Eeden, F. J., Cuppen, E., and Plasterk, R. H. (2003) The microRNA-producing enzyme Dicer1 is essential for zebrafish development. *Nat Genet* **35**, 217–218.
16. Rana, T. M. (2007) Illuminating the silence: Understanding the structure and function of small RNAs. *Nat Rev* **8**, 23–36.
17. Sharp, P. A., and Zamore, P. D. (2000) Molecular biology. RNA interference. *Science* **287**, 2431–2433.
18. Hammond, S. M., Boettcher, S., Caudy, A. A., Kobayashi, R., and Hannon, G. J. (2001) Argonaute2, a link between genetic and biochemical analyses of RNAi. *Science* **293**, 1146–1150.
19. Meister, G., and Tuschl, T. (2004) Mechanisms of gene silencing by double-stranded RNA. *Nature* **431**, 343–349.
20. Ishizuka, A., Siomi, M. C., and Siomi, H. (2002) A Drosophila fragile X protein interacts with components of RNAi and ribosomal proteins. *Genes Dev* **16**, 2497–2508.
21. Caudy, A. A., Ketting, R. F., Hammond, S. M., Denli, A. M., Bathoorn, A. M., Tops, B. B., Silva, J. M., Myers, M. M., Hannon, G. J., and Plasterk, R. H. (2003) A micrococcal nuclease homologue in RNAi effector complexes. *Nature* **425**, 411–414.
22. Mourelatos, Z., Dostie, J., Paushkin, S., Sharma, A., Charroux, B., Abel, L., Rappsilber, J., Mann, M., and Dreyfuss, G. (2002) miRNPs: A novel class of ribonucleoproteins containing numerous microRNAs. *Genes Dev* **16**, 720–728.
23. Dostie, J., Mourelatos, Z., Yang, M., Sharma, A., and Dreyfuss, G. (2003) Numerous microRNPs in neuronal cells containing novel microRNAs. *RNA* **9**, 180–186.
24. Nelson, P. T., Hatzigeorgiou, A. G., and Mourelatos, Z. (2004) miRNP: mRNA association in polyribosomes in a human neuronal cell line. *RNA* **10**, 387–394.
25. Chendrimada, T. P., Gregory, R. I., Kumaraswamy, E., Norman, J., Cooch, N., Nishikura, K., and Shiekhattar, R. (2005) TRBP recruits the Dicer complex to Ago2 for microRNA processing and gene silencing. *Nature* **436**, 740–744.
26. Lee, Y., Hur, I., Park, S. Y., Kim, Y. K., Suh, M. R., and Kim, V. N. (2006) The role of

PACT in the RNA silencing pathway. *EMBO J* **25**, 522–532.
27. Chu, C. Y., and Rana, T. M. (2006) Translation repression in human cells by microRNA-induced gene silencing requires RCK/p54. *PLoS Biol* **4**, e210.
28. Carmell, M. A., Xuan, Z., Zhang, M. Q., and Hannon, G. J. (2002) The Argonaute family: Tentacles that reach into RNAi, developmental control, stem cell maintenance, and tumorigenesis. *Genes Dev* **16**, 2733–2742.
29. Sasaki, T., Shiohama, A., Minoshima, S., and Shimizu, N. (2003) Identification of eight members of the Argonaute family in the human genome small star, filled. *Genomics* **82**, 323–330.
30. Williams, R. W., and Rubin, G. M. (2002) ARGONAUTE1 is required for efficient RNA interference in Drosophila embryos. *Proc Natl Acad Sci USA* **99**, 6889–6894.
31. Okamura, K., Ishizuka, A., Siomi, H., and Siomi, M. C. (2004) Distinct roles for Argonaute proteins in small RNA-directed RNA cleavage pathways. *Genes Dev* **18**, 1655–1666.
32. Siomi, H., Ishizuka, A., and Siomi, M. C. (2004) RNA interference: A new mechanism by which FMRP acts in the normal brain? What can Drosophila teach us?, *Ment Retard Dev Disabil Res Rev* **10**, 68–74.
33. Caudy, A. A., Myers, M., Hannon, G. J., and Hammond, S. M. (2002) Fragile X-related protein and VIG associate with the RNA interference machinery. *Genes Dev* **16**, 2491–2496.
34. Hammond, S. M., Bernstein, E., Beach, D., and Hannon, G. J. (2000) An RNA-directed nuclease mediates post-transcriptional gene silencing in Drosophila cells. *Nature* **404**, 293–296.
35. Elbashir, S. M., Lendeckel, W., and Tuschl, T. (2001) RNA interference is mediated by 21- and 22-nucleotide RNAs. *Genes Dev* **15**, 188–200.
36. Martinez, J., Patkaniowska, A., Urlaub, H., Luhrmann, R., and Tuschl, T. (2002) Single-stranded antisense siRNAs guide target RNA cleavage in RNAi. *Cell* **110**, 563–574.
37. Ambros, V. (2001) MicroRNAs: Tiny regulators with great potential. *Cell* **107**, 823–826.
38. Ambros, V. (2004) The functions of animal microRNAs. *Nature* **431**, 350–355.
39. Bartel, D. P. (2004) MicroRNAs: Genomics, biogenesis, mechanism, and function. *Cell* **116**, 281–297.
40. Carrington, J. C., and Ambros, V. (2003) Role of microRNAs in plant and animal development. *Science* **301**, 336–338.
41. Zhang, B. H., Stellwag, E. J., and Pan, X. P. (2009) Large-scale genome analysis reveals unique features of microRNAs. *Gene* **443**, 100–109.
42. Yeom, K. H., Lee, Y., Han, J., Suh, M. R., and Kim, V. N. (2006) Characterization of DGCR8/Pasha, the essential cofactor for Drosha in primary miRNA processing. *Nucleic Acids Res* **34**, 4622–4629.
43. Lund, E., Guttinger, S., Calado, A., Dahlberg, J. E., and Kutay, U. (2004) Nuclear export of microRNA precursors. *Science* **303**, 95–98.
44. Gregory, R. I., Chendrimada, T. P., Cooch, N., and Shiekhattar, R. (2005) Human RISC couples microRNA biogenesis and posttranscriptional gene silencing. *Cell* **123**, 631–640.
45. Gregory, R. I., Chendrimada, T. P., and Shiekhattar, R. (2006) MicroRNA biogenesis: Isolation and characterization of the microprocessor complex. *Methods Mol Biol* **342**, 33–47.
46. Eulalio, A., Behm-Ansmant, I., and Izaurralde, E. (2007) P bodies: At the crossroads of post-transcriptional pathways. *Nat Reviews* **8**, 9–22.
47. Lee, R. C., Feinbaum, R. L., and Ambros, V. (1993) The *C. elegans* heterochronic gene *lin-4* encodes small RNAs with antisense complementarity to *lin-14*. *Cell* **75**, 843–854.
48. Wightman, B., Ha, I., and Ruvkun, G. (1993) Posttranscriptional regulation of the heterochronic gene *Lin-14* by *Lin-4* mediates temporal pattern-formation in *C. elegans*. *Cell* **75**, 855–862.
49. Lee, R. C., and Ambros, V. (2001) An extensive class of small RNAs in *Caenorhabditis elegans*. *Science* **294**, 862–864.
50. Lagos-Quintana, M., Rauhut, R., Lendeckel, W., and Tuschl, T. (2001) Identification of novel genes coding for small expressed RNAs. *Science* **294**, 853–858.
51. Lau, N. C., Lim, L. P., Weinstein, E. G., and Bartel, D. P. (2001) An abundant class of tiny RNAs with probable regulatory roles in *Caenorhabditis elegans*. *Science* **294**, 858–862.
52. Reinhart, B. J., Slack, F. J., Basson, M., Pasquinelli, A. E., Bettinger, J. C., Rougvie, A. E., Horvitz, H. R., and Ruvkun, G. (2000) The 21-nucleotide let-7 RNA regulates developmental timing in *Caenorhabditis elegans*. *Nature* **403**, 901–906.
53. Chen, J. F., Mandel, E. M., Thomson, J. M., Wu, Q., Callis, T. E., Hammond, S. M., Conlon, F. L., and Wang, D. Z. (2006) The role of microRNA-1 and microRNA-133 in

skeletal muscle proliferation and differentiation. *Nat Genet* **38**, 228–233.
54. Brennecke, J., Hipfner, D. R., Stark, A., Russell, R. B., and Cohen, S. M. (2003) bantam encodes a developmentally regulated microRNA that controls cell proliferation and regulates the proapoptotic gene hid in Drosophila. *Cell* **113**, 25–36.
55. Xu, P., Vernooy, S. Y., Guo, M., and Hay, B. A. (2003) The Drosophila microRNA Mir-14 suppresses cell death and is required for normal fat metabolism. *Curr Biol* **13**, 790–795.
56. Baroukh, N., Ravier, M. A., Loder, M. K., Hill, E. V., Bounacer, A., Scharfmann, R., Rutter, G. A., and Van Obberghen, E. (2007) MicroRNA-124a regulates Foxa2 expression and intracellular signaling in pancreatic beta-cell lines. *J Biol Chem* **282**, 19575–19588.
57. Webster, R. J., Giles, K. M., Price, K. J., Zhang, P. M., Mattick, J. S., and Leedman, P. J. (2009) Regulation of epidermal growth factor receptor signaling in human cancer cells by microRNA-7. *J Biol Chem* **284**, 5731–5741.
58. Zhang, B. H., Pan, X. P., Cobb, G. P., and Anderson, T. A. (2007) MicroRNAs as oncogenes and tumor suppressors. *Dev Biol* **302**, 1–12.
59. Saito, Y., Liang, G., Egger, G., Friedman, J. M., Chuang, J. C., Coetzee, G. A., and Jones, P. A. (2006) Specific activation of microRNA-127 with downregulation of the proto-oncogene BCL6 by chromatin-modifying drugs in human cancer cells. *Cancer Cell* **9**, 435–443.
60. Pearson, J. C., Lemons, D., and McGinnis, W. (2005) Modulating *Hox* gene functions during animal body patterning. *Nat Rev Genet* **6**, 893–904.
61. Hammond, S. M. (2006) MicroRNAs as oncogenes. *Curr Opin Genet Dev* **16**, 4–9.
62. Hammond, S. M. (2006) MicroRNAs as oncogenes. *Curr Opin Genet Dev* **16**, 4–9.
63. Kosik, K. S. (2006) The neuronal microRNA system. *Nat Rev Neurosci* **7**, 911–920.
64. Kim, J., Krichevsky, A., Grad, Y., Hayes, G. D., Kosik, K. S., Church, G. M., and Ruvkun, G. (2004) Identification of many microRNAs that copurify with polyribosomes in mammalian neurons. *Proc Natl Acad Sci USA* **101**, 360–365.
65. Krichevsky, A. M., King, K. S., Donahue, C. P., Khrapko, K., and Kosik, K. S. (2003) A microRNA array reveals extensive regulation of microRNAs during brain development. *RNA* **9**, 1274–1281.
66. Giraldez, A. J., Cinalli, R. M., Glasner, M. E., Enright, A. J., Thomson, J. M., Baskerville, S., Hammond, S. M., Bartel, D. P., and Schier, A. F. (2005) MicroRNAs regulate brain morphogenesis in zebrafish. *Science* **308**, 833–838.
67. Miska, E. A., Alvarez-Saavedra, E., Townsend, M., Yoshii, A., Sestan, N., Rakic, P., Constantine-Paton, M., and Horvitz, H. R. (2004) Microarray analysis of microRNA expression in the developing mammalian brain. *Genome Biol* **5**, R68.
68. Sempere, L. F., Freemantle, S., Pitha-Rowe, I., Moss, E., Dmitrovsky, E., and Ambros, V. (2004) Expression profiling of mammalian microRNAs uncovers a subset of brain-expressed microRNAs with possible roles in murine and human neuronal differentiation. *Genome Biol* **5**, R13.
69. Ashraf, S. I., and Kunes, S. (2006) A trace of silence: Memory and microRNA at the synapse. *Curr Opin Neurobiol* **16**, 535–539.
70. Zhang, B. H., Pan, X. P., and Anderson, T. A. (2006) MicroRNA: A new player in stem cells. *J Cell Physiol* **209**, 266–269.
71. Forstemann, K., Tomari, Y., Du, T. T., Vagin, V. V., Denli, A. M., Bratu, D. P., Klattenhoff, C., Theurkauf, W. E., and Zamore, P. D. (2005) Normal microRNA maturation and germ-line stem cell maintenance requires loquacious, a double-stranded RNA- binding domain protein. *PLoS Biol* **3**, 1187–1201.
72. Gangaraju, V. K., and Lin, H. F. (2009) MicroRNAs: Key regulators of stem cells. *Nat Rev Mol Cell Biol* **10**, 116–125.
73. Greco, S. J., and Rameshwar, P. (2007) microRNAs regulate synthesis of the neurotransmitter substance P in human mesenchymal stem cell-derived neuronal cells. *Proc Natl Acad Sci USA* **104**, 15484–15489.
74. Hammond, S. M., and Sharpless, N. E. (2008) HMGA2, microRNAs, and stem cell aging. *Cell* **135**, 1013–1016.
75. Hatfield, S., and Ruohola-Baker, H. (2008) microRNA and stem cell function. *Cell Tissue Res* **331**, 57–66.
76. Hatfield, S. D., Shcherbata, H. R., Fischer, K. A., Nakahara, K., Carthew, R. W., and Ruohola-Baker, H. (2005) Stem cell division is regulated by the microRNA pathway. *Nature* **435**, 974–978.
77. Houbaviy, H. B., Murray, M. F., and Sharp, P. A. (2003) Embryonic stem cell-specific microRNAs. *Dev Cell* **5**, 351–358.
78. Ivey, K. N., Muth, A., Arnold, J., King, F. W., Yeh, R. F., Fish, J. E., Hsiao, E. C., Schwartz, R. J., Conklin, B. R., Bernstein, H. S., and Srivastava, D. (2008) microRNA regulation

of cell lineages in mouse and human embryonic stem cells. *Cell Stem Cell* **2**, 219–229.
79. Kanellopoulou, C., Muljo, S. A., Kung, A. L., Ganesan, S., Drapkin, R., Jenuwein, T., Livingston, D. M., and Rajewsky, K. (2005) Dicer-deficient mouse embryonic stem cells are defective in differentiation and centromeric silencing. *Genes Dev* **19**, 489–501.
80. Krichevsky, A. M., Sonntag, K. C., Isacson, O., and Kosik, K. S. (2006) Specific microRNAs modulate embryonic stem cell-derived neurogenesis. *Stem Cells* **24**, 857–864.
81. Kuwabara, T., Hsieh, J., Nakashima, K., Taira, K., and Gage, F. H. (2004) A small modulatory dsRNA specifies the fate of adult neural stem cells. *Cell* **116**, 779–793.
82. Lakshmipathy, U., Love, B., Goff, L. A., Jornsten, R., Graichen, R., Hart, R. P., and Chesnut, J. D. (2007) microRNA expression pattern of undifferentiated and differentiated human embryonic stem cells. *Stem Cells Dev* **16**, 1003–1016.
83. Li, Q. T., and Gregory, R. I. (2008) microRNA regulation of stem cell fate. *Cell Stem Cell* **2**, 195–196.
84. Murchison, E. P., Partridge, J. F., Tam, O. H., Cheloufi, S., and Hannon, G. J. (2005) Characterization of Dicer-deficient murine embryonic stem cells. *Proc Natl Acad Sci USA* **102**, 12135–12140.
85. Park, J. K., Liu, X., Strauss, T. J., McKearin, D. M., and Liu, Q. H. (2007) The miRNA pathway intrinsically controls self-renewal of Drosophila germline stem cells. *Curr Biol* **17**, 533–538.
86. Suh, M. R., Lee, Y., Kim, J. Y., Kim, S. K., Moon, S. H., Lee, J. Y., Cha, K. Y., Chung, H. M., Yoon, H. S., Moon, S. Y., Kim, V. N., and Kim, K. S. (2004) Human embryonic stem cells express a unique set of microRNAs. *Dev Biol* **270**, 488–498.
87. Tang, F. C., Hajkova, P., Barton, S. C., Lao, K. Q., and Surani, M. A. (2006) microRNA expression profiling of single whole embryonic stem cells. *Nucleic Acids Res* **34**, e9.
88. Wang, Y., Baskerville, S., Shenoy, A., Babiarz, J. E., Baehner, L., and Blelloch, R. (2008) Embryonic stem cell-specific microRNAs regulate the G1-S transition and promote rapid proliferation. *Nat Genet* **40**, 1478–1483.
89. Wang, Y. L., Keys, D. N., Au-Young, J. K., and Chen, C. F. (2009) microRNAs in embryonic stem cells. *J Cell Physiol* **218**, 251–255.
90. Zou, G. M., and Yoder, M. C. (2005) Application of RNA interference to study stem cell function: Current status and future perspectives. *Biol Cell* **97**, 211–219.
91. Marson, A., Levine, S. S., Cole, M. F., Frampton, G. M., Brambrink, T., Johnstone, S., Guenther, M. G., Johnston, W. K., Wernig, M., Newman, J., Calabrese, J. M., Dennis, L. M., Volkert, T. L., Gupta, S., Love, J., Hannett, N., Sharp, P. A., Bartel, D. P., Jaenisch, R., and Young, R. A. (2008) Connecting microRNA genes to the core transcriptional regulatory circuitry of embryonic stem cells. *Cell* **134**, 521–533.
92. Mineno, J., Okamoto, S., Ando, T., Sato, M., Chono, H., Izu, H., Takayama, M., Asada, K., Mirochnitchenko, O., Inouye, M., and Kato, I. (2006) The expression profile of microRNAs in mouse embryos. *Nucleic Acids Res* **34**, 1765–1771.
93. Bernstein, E., Kim, S. Y., Carmell, M. A., Murchison, E. P., Alcorn, H., Li, M. Z., Mills, A. A., Elledge, S. J., Anderson, K. V., and Hannon, G. J. (2003) Dicer is essential for mouse development. *Nat Genet* **35**, 215–217.
94. Wang, Y., Medvid, R., Melton, C., Jaenisch, R., and Blelloch, R. (2007) DGCR8 is essential for microRNA biogenesis and silencing of embryonic stem cell self-renewal. *Nat Genet* **39**, 380–385.
95. Stefani, G., and Slack, F. J. (2008) Small non-coding RNAs in animal development. *Nat Rev* **9**, 219–230.
96. Sinkkonen, L., Hugenschmidt, T., Berninger, P., Gaidatzis, D., Mohn, F., Artus-Revel, C. G., Zavolan, M., Svoboda, P., and Filipowicz, W. (2008) microRNAs control de novo DNA methylation through regulation of transcriptional repressors in mouse embryonic stem cells. *Nat Struct Mol Biol* **15**, 259–267.
97. Velkey, J. M., and O'Shea, K. S. (2003) Oct4 RNA interference induces trophectoderm differentiation in mouse embryonic stem cells. *Genesis* **37**, 18–24.
98. Hay, D. C., Sutherland, L., Clark, J., and Burdon, T. (2004) Oct-4 knockdown induces similar patterns of endoderm and trophoblast differentiation markers in human and mouse embryonic stem cells. *Stem Cells* **22**, 225–235.
99. Ui-Tei, K., Zenno, S., Miyata, Y., and Saigo, K. (2000) Sensitive assay of RNA interference in Drosophila and Chinese hamster cultured cells using firefly luciferase gene as target. *FEBS Lett* **479**, 79–82.
100. Gil, J., and Esteban, M. (2000) Induction of apoptosis by the dsRNA-dependent protein kinase (PKR): Mechanism of action. *Apoptosis* **5**, 107–114.

101. Caplen, N. J., and Mousses, S. (2003) Short interfering RNA (siRNA)-mediated RNA interference (RNAi) in human cells. *Ann N Y Acad Sci* **1002**, 56–62.
102. Michaelson, J. S., and Leder, P. (2003) RNAi reveals anti-apoptotic and transcriptionally repressive activities of DAXX. *J Cell Sci* **116**, 345–352.
103. Brummelkamp, T. R., Bernards, R., and Agami, R. (2002) Stable suppression of tumorigenicity by virus-mediated RNA interference. *Cancer Cell* **2**, 243–247.

Section II

RNA Interference

Chapter 3

Forward RNAi Screens in Human Stem Cells

Christine Karlsson, Jonas Larsson, and Aurélie Baudet

Abstract

Identifying the genes and pathways that regulate self-renewal and differentiation in somatic stem cells is a central goal in stem cell and cancer biology. Here, we describe a method for RNAi-based screens in primary human hematopoietic stem and progenitor cells. These cells are suitable targets for complex, selection-based screens using pooled lentiviral shRNA libraries. The screening approach is a promising new tool to dissect regulatory mechanisms in hematopoietic and somatic stem cells, in general, and may be particularly useful to identify gene targets and modifiers that can be further exploited in strategies for ex vivo stem cell expansion.

Key words: Hematopoietic stem cells, RNA interference, forward genetic screens, positive selection, colony-forming cell assay.

1. Introduction

Stem cells harbor enormous proliferation and differentiation potential. This is evident not only during embryonic development and organogenesis, but also in the adult organism during tissue regeneration following injury. The clinical use of stem cells in regenerative medicine is, however, hampered by insufficient knowledge about the regulatory mechanisms governing key stem cell fate decisions such as self-renewal and differentiation. The defining ability of stem cells to reproduce and self-renew can be elegantly modeled in the mammalian hematopoietic system both in vitro and in vivo. Therefore, hematopoietic stem cells may serve as a paradigm to understand regulation of somatic stem cells in general.

B. Zhang, E.J. Stellwag (eds.), *RNAi and MicroRNA-Mediated Gene Regulation in Stem Cells*, Methods in Molecular Biology 650, DOI 10.1007/978-1-60761-769-3_3, © Springer Science+Business Media, LLC 2010

Genome-wide genetic screens to identify novel genes involved in specific biological processes were, until recently, limited to invertebrates such as worms and fruit flies. However, development of RNA interference (RNAi) technology and the very recent creation of large, short hairpin RNA (shRNA), retroviral and lentiviral vector libraries now provide tools to perform broad forward genetic screens by simultaneously silencing thousands of genes in primary mammalian cells (1–4). Complex mixtures of shRNAs can be assayed simultaneously in pools using selection-based screening approaches that filter cell populations based on phenotypic criteria, such as proliferation. The success of such screens is largely dependent on the ability of shRNA vectors to infect sufficient numbers of target cells as well as the specificity and sensitivity of assays to read out an altered phenotype. Human hematopoietic stem and progenitor cells (HSPCs) have clear advantages in both respects since they can be readily transduced in large numbers with lentiviral vectors, while retaining their immature characteristics, and further tested in highly accurate quantitative and qualitative assays for stem and progenitor cell function (5). When cultured without support from stroma cells, HSPCs can maintain an undifferentiated state only for a short time. This limited ability of HSPCs to sustain their immature properties under ex vivo culture conditions can be used as a basis for pooled screens where augmented proliferation capacity of shRNA-modified clones is detected by positive selection (**Fig. 3.1**). Applying this approach, we have identified novel modulators of human HSPCs, thereby demonstrating the potential of this forward genetic screening approach in primary human stem cell populations (6).

2. Materials

2.1. Screen

2.1.1. Isolation of CD34 Cells from Umbilical Cord Blood

1. Collecting medium: Dulbecco's Modified Eagle's Medium (DMEM) supplemented with 1% fetal bovine serum (FBS), 20 units/mL penicillin + 20 mg/mL streptomycin (PS), and 22 units/mL of heparin.
2. 75 cm^2nonventilated flasks.
3. Lymphoprep, $d = 1{,}077$ g/L (Axis Shield PoC AS).
4. DMEM supplemented with 10% FBS and PS.
5. Direct CD34 progenitor cell-isolation kit, LS columns, and Midi MACS magnetic holder (Miltenyi Biotec).
6. Isolation buffer: 2% FBS, 2 mM EDTA, pH 8, in PBS.

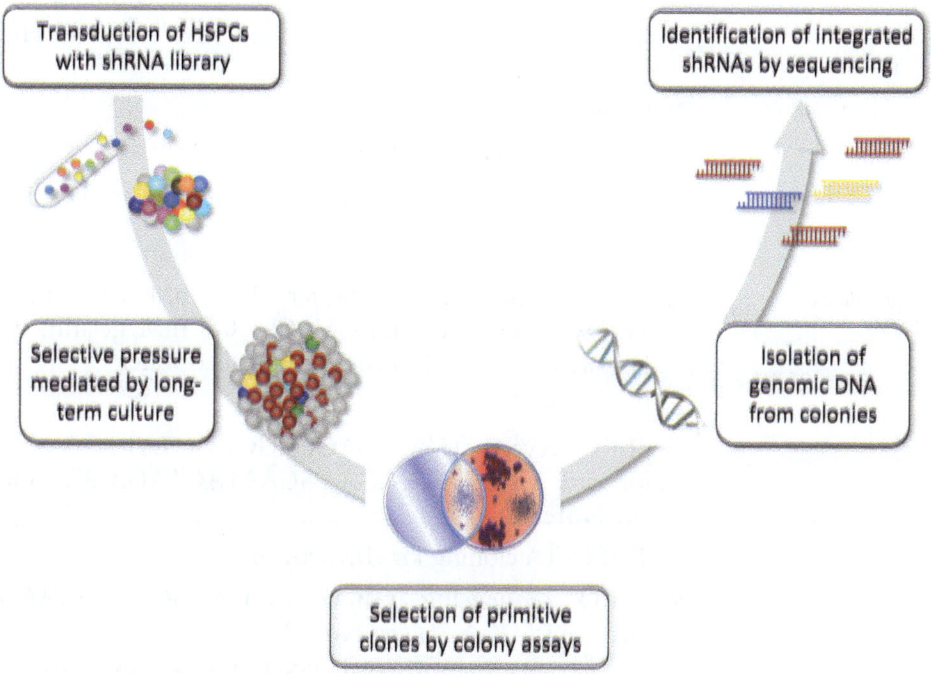

Fig. 3.1. RNA interference screen in human hematopoietic stem cells. Large numbers of primary umbilical cord blood-derived human HSPCs are infected with lentiviral shRNA libraries and subsequently passaged in long-term cultures. After long-term cultivation, CFC assays are performed to positively select for clones that have acquired enhanced self-renewal/proliferation ability. Potential hits are identified by sequence analysis of proviral inserts from the selected cells.

7. Freezing medium A: DMEM supplemented with 20% FBS and PS.
8. Freezing medium B: Freezing medium A supplemented with 20% dimethylsulfoxide.
9. Trypan Blue solution.

2.1.2. Lentiviral Transduction and Culture of CD34+ Cells

1. StemSpan serum-free expansion medium for hematopoietic cells SFEM (StemCell Technologies) supplemented with PS.
2. Human stem cell factor (SCF), FLT3 ligand (FL), and Thrombopoietin (TPO).
3. 48-multiwell plate. Nontreated polystyrene.
4. 40 μg/mL RetroNectin solution in PBS (Takara Bio Inc.).
5. 2% bovine serum albumin solution (BSA) in PBS.
6. Pooled lentiviral shRNA library, e.g., The RNAi Consortium (TRC), supplied as MISSION shRNA library by Sigma Aldrich (*see* **Note 1**). Handle lentiviral particles according to the recommendations for risk group level 2 (RGL-2).

2.1.3. Colony-Forming-Cell (CFC) Assay

1. Methocult methylcellulose medium with cytokines H4434 (StemCell Technologies).
2. Puromycin.
3. 6-well tissue culture plates.
4. Syringes (2 and 10 mL) and blunt-end needles (16 × 11/2″).

2.1.4. Identification of Target shRNAs

1. 0.22-mm filtered lysis buffer: 105 mM KCl, 14 mM TrisHCl, pH 8, 2.5 mM $MgCl_2$, 0.3 mM gelatin, 0.45% Igepal, and 0.45% Tween-20 in distilled water.
2. Proteinase K.
3. pLKO reverse (5′-CTGAGCAGCCGCTATTGG-3′) and forward (5′-AAACCCGAAGGAATAGAAGA-3′) primers (*see* **Note 1**).
4. TOPO-TA cloning kit (Invitrogen).
5. pLKO sequencing primer (5′-CAAGGCTGTTAGAGA GATAATTGGA-3′) (*see* **Note 1**).
6. Kit for isolation of genomic DNA (DNeasy, Qiagen).

2.2. Validation of shRNAs

2.2.1. Subcloning of shRNAs

1. pLKO1 vector (*see* **Note 1**).
2. Restriction enzymes NdeI and SpeI (New England Biolabs, NEB).
3. Gel extraction kit (Qiagen).
4. Nucleotide removal kit (Qiagen).
5. T4 DNA ligase and ligase buffer.
6. Chemo-competent XL10 gold bacteria (Stratagene) or another $RecA^-$ strain.
7. LB and SOC media, ampicillin.
8. Plasmid purification kit (Qiagen).

2.2.2. Virus Production and Titration

1. Human 293T cells (DMSZ, German Collection of Microorganisms and Cell Cultures).
2. Packaging plasmid pCMVΔR8.91, envelope plasmid VSV-G pMDG (http://www.addgene.org/Didier_Trono), and HIV plasmid containing the shRNA construct.
3. 2× Hepes buffer solution (HeBS), sterile filtered, pH 7.0: 50 mM Hepes, 280 mM NaCl, 1.5 mM Na_2HPO_4 dissolved in 1,000 mL double-distilled H_2O and stored at –20°C in 50 mL aliquots.
4. 0.5 M $CaCl_2$, stored at –80°C.

5. Water for mixing plasmids (p-H_2O): 50 mL sterile H_2O, 125 μL Hepes 1 M pH 7.3. Stored at 4°C for no longer than 1 week.
6. 0.45 μm Millipore Express PLUS (PES) Membrane Filter, low clogging, low protein-binding (Millipore).
7. Ultracentrifuge (Beckman Coulter, Optima™).

3. Methods

The basis of the described approach is to target large numbers of human HSPCs with pooled lentiviral shRNA libraries and use their limited self-renewal capacity in vitro as a basis for positive selection of clones that have gained enhanced self-renewal/proliferation properties. Library-transduced cells are passaged in long-term liquid cultures (12 weeks) and assayed in colony-forming cell (CFC) assays allowing positive selection of clones with sustained primitive potential. Following selection, the colonies from each pool are harvested for DNA extraction and subsequent PCR amplification of the proviral inserts. The PCR products are then sequenced to determine the distribution of shRNAs within each pool. Abundant shRNAs represent candidate modifiers of self-renewal and proliferation in HSPCs.

To functionally validate the outcome of the screen, individual candidate shRNAs are subcloned into the lentiviral vector and re-introduced into HSPCs. shRNAs showing a positive score in the validation experiment are further examined for their target specificity (**Fig. 3.2**). For that purpose, the capacity to knockdown target gene expression is evaluated by quantitative real-time PCR. If sufficient suppression of target transcript levels is achieved with the shRNA construct retrieved from the screen, additional hairpins targeting the gene of interest are tested to further confirm the target specificity.

3.1. Screen

3.1.1. Isolation of CD34+ Cells from Umbilical Cord Blood

1. Human hematopoietic stem and progenitor cells (HSPCs) are isolated from umbilical cord blood based on expression of the cell surface marker CD34 (*see* **Note 2**). Collect cord blood samples in 75 cm^2 nonventilated flasks containing 20 mL of collecting medium. Keep at 4°C until processing.
2. Add Lymphoprep in 50-mL Falcon tubes as ½ the volume of blood (up to 15 mL of Lymphoprep for 35 mL of blood). Gently add blood on top of Lymphoprep. Blood and Lymphoprep must not mix: keep the interface thin and

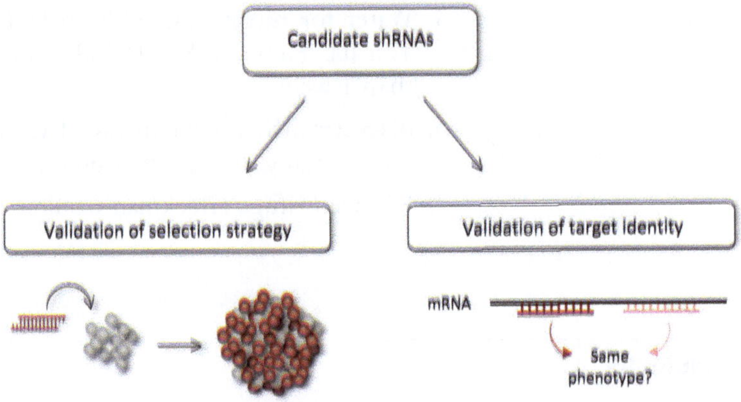

Fig. 3.2. Candidate shRNA validation. The validation of the shRNA screen has two components: candidate shRNAs are first individually tested for their ability to reproduce the functional outcome from the screen (expansion of progenitor cells); and gene target specificity is validated by assessing knockdown efficiency and by testing additional shRNAs targeting the same gene.

even. Drops and high flow should be avoided. Centrifuge for 20 min at RT, 850×*g* with minimal break.

3. Using a plastic Pasteur pipette with reservoir, transfer the ring of mononuclear cells to a 50-mL tube, taking out as few red cells as possible. Dilute the sample at least twice with DMEM-10% FCS. Centrifuge for 5 min at 400×*g*, with normal break.
4. Re-suspend cells in 10 mL of isolation buffer, filter suspension through a 50 μM cell strainer and transfer to a 15-mL tube. Prepare a 10-fold dilution and count cells using a hemocytometer after Trypan Blue staining.
5. Centrifuge the cell suspension for 5 min at 350×*g*. Carefully aspirate the supernatant. Re-suspend sedimented cells in 300 μL isolation buffer (up to 300 million cells per 300 μL), add 100 μL of blocking reagent, and mix well by pipetting up and down a few times. Add 100 μL of a suspension of CD34 microbeads to the cell suspension and mix well again.
7. Incubate for 30 min at 4°C: Mix every 10 min by tilting the tube from side to side a few times.
8. Add 10 mL of isolation buffer, and centrifuge for 5 min at 300×*g*. Re-suspend cells in isolation buffer (up to 300 million cells in 500 μL buffer). Magnetic microbeads are now bound to the $CD34^+$ cells.
9. Place a MidiMACS column on magnetic holder and equilibrate it with 3 mL of isolation buffer. Add 500 μL of cell suspension to the column and wash three times with 3 mL

of isolation buffer. The $CD34^+$ cells are now attached to the column by magnetic force.

10. Remove the column from the magnetic holder and place it over a 15 mL Falcon collection tube, add 5 mL of isolation buffer, and elute the cells from the column into the collection tube using the supplied plunger. Centrifuge the collection tube containing the eluted cells for 5 min at 300 × *g* Re-suspend the cells in 500 μL isolation buffer and repeat the purification process once with a new column according to the procedure outlined in Step 9.
11. To freeze $CD34^+$ cells, re-suspend cells at room temperature in 2× freezing medium A (250 μL for 1 to 5 × 10^6 cells). Add an equal volume of room temperature 2× freezing medium B after the addition of medium A. Place the tube containing the suspended cells in a freezing container at room temperature and transfer immediately to −80°C. Store at −80°C for 24 h and transfer to liquid nitrogen tank for longer storage.

3.1.2. Library Transduction and Long-Term Culture of $CD34^+$ Cells

Day 1

1. Plate freshly isolated or thawed $CD34^+$ cells in a tissue culture-treated multiwell plate or small flask at a density of 0.75 × 10^6/mL in fresh culture medium (SFEM supplemented with 100 ng/mL SCF, 100 ng/mL TPO, and 100 ng/mL FL) and incubate overnight at 37°C in an atmosphere of 5% CO_2.

Day 2

1. Coat appropriate number of wells (*see* **Note 3**) in nontissue-culture 48-multiwell plate with 40 μg/mL RetroNectin solution (150 μL/well) at room temperature for 2 h. Include wells for nontransduced cells (Mock) and control-shRNA transduced cells (*see* **Note 3**).
2. Remove RetroNectin and block with 2% BSA solution in PBS (200 μL/well) for 30 min at room temperature.
3. Remove BSA solution and add 2 × 10^5 $CD34^+$ cells per well in 0.3 mL medium.
4. Add lentiviral particles at a multiplicity of infection (MOI) resulting in approximately 35% transduction efficiency (*see* **Note 4**) to the wells and incubate overnight at 37°C in an atmosphere of 5% CO_2.

Day 3

1. Remove half of the medium and replace with fresh medium.

Day 4

1. Remove half of the medium and transfer the cells to 12-well plates in 1 mL expansion medium.

2. To evaluate transduction efficiency, plate CFCs for three to four pools in puromycin-containing (2 μg/mL) and puromycin-free medium as described in **Section 3.1.3**. The transduction percentage can be calculated according to the following formula: (CFC in puromycin medium/CFC in puromycin-free medium) × 100.

Day 5 and On

1. The cells should now be cultured long-term (up to 12 weeks) in expansion medium. The $CD34^+$ cells will divide actively during the first two weeks and will then divide more slowly. When confluent, the cells can be transferred from the 48-well plates to 12-well plates. Thereafter, continue to passage the cells in the 12-well plates throughout the long-term culture. When the cells become confluent, remove and discard 2/3 of the cells and replace with fresh medium.
2. To assess the proportion of primitive cells (i.e., progenitor cells with colony-forming ability), perform CFC assays for all the screening pools and controls after 6, 8, 10, and 12 weeks of culture.

3.1.3. CFC Assay

1. Supplement half of the methylcelluose medium with puromycin (2 μg/mL). Vortex and let stand to remove bubbles.
2. Using a syringe and needle, aliquot the CFC medium in FACS tubes, 1 mL per tube and note whether the tubes contain puromycin or not.
3. Add cells: 500 and 2,000 $CD34^+$ cells per pool to both puromycin-containing and puromycin-free medium. Vortex at maximum frequency for few seconds and let stand to remove bubbles.
4. Using a syringe and needle, add the cell suspension to a 35-mm diameter dish (or a well from a 6-well plate).
5. Spread the medium all over the well by tilting and turning the dish.
6. Score colonies after 14 days. *See* **Fig. 3.3** for examples of different colony types. Pools of interest are identified by one or more of the following criteria: (i) Overall higher colony numbers than the controls; (ii) Increased frequency of library-transduced puromycin-resistant colonies; and (iii) High frequency of rare colony types such as Burst Forming Unit Erythroid (BFU-E) colonies (**Fig. 3.3**).

3.1.4. Identification of Candidate shRNAs

1. The colonies from screening pools showing enhanced growth of progenitor cells at any of the time points analyzed (see criteria above) should be harvested to identify the most abundant shRNAs in those pools by PCR amplification

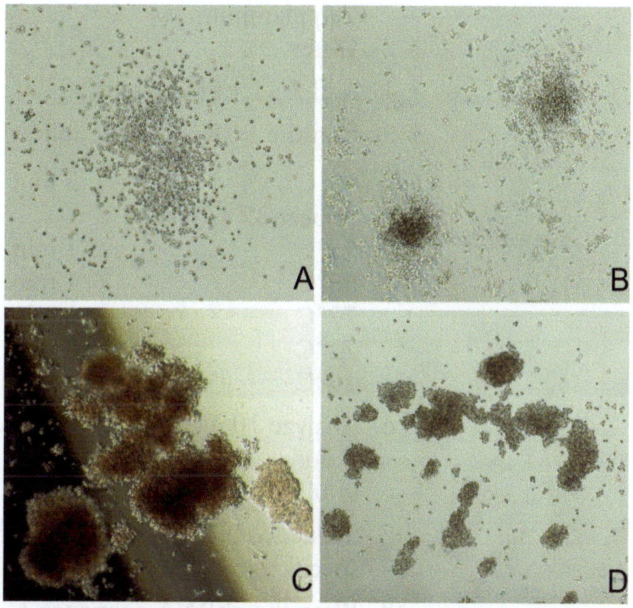

Fig. 3.3. Human hematopoietic colonies from cord blood CD34^{+} cells. Fully mature colonies develop in 14 days. (**A**) Colony-forming unit macrophage (CFU-M), (**B**) Colony-forming unit granulocyte/macrophage (CFU-GM), (**C**) Burst-forming unit erythroid (BFU-E), (**D**) Colony-forming unit granulocyte (CFU-G).

and sequencing of the proviral inserts. To enable clonal sequencing of shRNAs, two alternative approaches are possible. Either individual colonies from the methylcellulose plates are picked and analyzed or colonies are harvested in bulk from each pool and the PCR products are cloned in bacteria before sequencing.

Alternative 1

1. While observing the culture under an inverted microscope and using a micropipette set on 2 μL, aspirate an individual colony. Transfer to a PCR tube prefilled with 25 mL of lysis buffer supplemented with 60 mg/mL proteinase K. Pick 10–20 colonies per pool.
2. Incubate for 1 h at 50°C to lyse the sample and 15 min at 95°C to inactivate the proteinase K. Freeze or proceed with the PCR.
3. In a DNA-free environment, mix (applicable to TRC vector pLKO1, *see* **Note 1**):

Reverse primer (10 mM)	1 μL
Forward primer (10 mM)	1 μL
10× Platinium Taq Buffer	0.5 μL
Lysis buffer	3 μL
$MgCl_2$ (50 mM)	0.375 μL
dNTP 10 mM	0.5 μL

Taq platinium	0.25 μL
Water	11.375 μL

4. Add 7 μL of the lysate.
5. Run the PCR using the following program:

95°C	5 min
Repeat 32 cycles:	
95°C	1 min
55°C	1 min
72°C	1 min
72°C	8 min
10°C	end

6. Analyze 10 μL of the PCR products on a 1.2% agarose-TAE 1× gel. The expected size of the amplification product is 900 bp (*see* **Note 1**).
7. Sequence PCR products from positive samples using library-specific sequencing primer.
8. Blast the shRNA sequence to human transcriptome using Blast at http://blast.ncbi.nlm.nih.gov or use shRNA search tools provided by the library manufacturer to identify the shRNA sequence.

Alternative 2

1. Re-suspend and harvest colonies from each well by adding 2 mL of PBS to the well and pellet by centrifuging for 5 min at 300 × g at 4°C.
2. Extract DNA using Genomic DNA extraction kit. Quantify the DNA by measuring absorbance on a spectrophotometer and freeze DNA or proceed with the PCR.
3. In a DNA-free environment, mix (applicable to TRC vector pLKO1, *see* **Note 1**):

Reverse primer (10 mM)	1.5 μL
Forward primer (10 mM)	1.5 μL
10× Platinium Taq buffer	2.5 μL
$MgCl_2$ (50 mM)	0.75 μL
dNTP (10 mM)	1 μL
Taq platinium	0.15 μL
DNA	100 ng
Water up to	25 μL

4. PCR program: refer to *alternative 1.*
5. Analyze 10 μL of the PCR product on a 1% agarose-TAE 1× gel. The expected size of the amplification product is 900 bp (applicable to TRC vector pLKO1, *see* **Note 1**).
6. TOPO-TA-clone PCR products from positive samples according to kit manufacturer's instructions.

7. PCR amplify the insert from bacterial clones using TA-vector-specific primers and conditions according to kit manufacturer's instructions.
8. Sequence PCR products from positive samples using the library-specific sequencing primer.
9. Blast the shRNA sequence to human transcriptome with Blast at http://blast.ncbi.nlm.nih.gov or use shRNA search tools provided by the library manufacturer to identify the shRNA sequence.

3.2. Validation of Candidate shRNAs

Among the shRNAs detected by sequencing, the most prominent hits are selected for individual validation. As a rule of thumb, shRNAs showing a frequency of more than 30% in a given pool or sequences that are detected in more than one screening pool can be considered interesting candidates.

3.2.1. Cloning of Individual shRNAs into Lentiviral Library Vector

1. Digest 20 µg of TRC pLKO1 vector with NdeI and SpeI (*see* **Note 1**). Our suggestion is to use NEB enzymes since this system is compatible with a one-step reaction.
2. Analyze the digestion products on a 1% agarose-TAE gel. The expected sizes for the digestion products are 600 bp and 7.1 kb. Extract the 7.1 kb band using gel extraction kit. Quantify DNA by measuring absorbance on a spectrophotometer.
3. Purify the clonal shRNA-containing PCR products from **Section 3.1.4** with a nucleotide removal kit; digest the PCR products with NdeI and SpeI.
4. Analyze the digestion products on a 1.2% agarose-TAE gel. Extract the 600 pb band using gel extraction kit. Quantify DNA by measuring absorbance on a spectrophotometer.
5. Proceed to ligation:

NdeI/SpeI-digested pLKO1 vector	200 ng
NdeI/SpeI-digested PCR product	50 ng
10× T4 DNA ligase buffer	2 µL
T4 DNA ligase	0.5 units
Water up to	20 µL

6. Incubate for 1 h at 25°C and 15 min at 65°C. Freeze or proceed with transformation.
7. Thaw an aliquot of chemo-competent RecA^{-} bacteria on ice.
8. Add 10 µL of the ligation. Mix by tilting the tube. Incubate for 30 min on ice.

9. Transform bacteria and grow on LB agar ampicillin plates. Pick colonies and purify the plasmid using plasmid purification kit.
10. Sequence with the library-specific sequencing primer to verify the identity of the shRNA.

3.2.2. Virus Production

1. Viruses are produced in 10-cm culture dishes in culture medium consisting of DMEM supplemented with 10% FBS and P/S.
2. Plate human 293T cells in culture medium. Adjust the cell number to achieve a culture density equivalent to 50–70% of confluency on the day of transfection.
3. Aspirate and add new culture medium 2–3 h prior to transfection.
4. Mix packaging plasmid pCMVΔR8.91 (14 μg), envelope plasmid VSV-G pMDG (6 μg), and plasmid DNA containing the shRNA construct (20 μg) in a suitable tube and add d-H_2O to a volume of 250 μL.
5. Add 250 μL of 0.5 M $CaCl_2$ solution.
6. Vortex the tube while adding 0.5 mL HeBS. Add the DNA solution dropwise over 2–4 min.
7. Incubate at room temperature for 20–25 min to allow precipitate to form. The mixture will appear hazy.
8. Distribute mixture dropwise and evenly over the plate containing 293T cells and mix gently.
9. Incubate overnight (18 h) at 37°C in an atmosphere of 5% CO_2.
10. Aspirate medium carefully avoiding detachment of the cell layer. The medium needs to be removed completely to avoid the precipitate in the viral supernatant.
11. Gently add 6 mL of culture medium and incubate at 33°C in an atmosphere of 5% CO_2.
12. Harvest the supernatant 48 and 72 h posttransfection through filtration using a 0.45-μm PES filter.
13. Transfer the filtered supernatant to an ultracentrifuge tube and centrifuge at 75,600×*g* for 90 min at 4°C.
14. Gently aspirate medium (a white translucent pellet should be barely visible), refill tube with supernatant and centrifuge again at 75,600×*g* for 90 min at 4°C. The procedure can be repeated at most three times before a new ultracentrifuge tube is used.
15. After the final ultracentrifugation, aspirate supernatant and invert tubes on absorbent paper for 5 min.

16. Add 75–300 μL of SFEM medium to the tube and resuspend. Seal the tube with laboratory film and incubate at room temperature for 30 min to allow dissolving of the pellet.
17. Move the viral concentrate to a screw-cap microcentrifuge tube and wash the ultracentrifuge tube with an additional 100 μL of SFEM medium.
18. Vortex (maximum speed) the concentrated virus for 5 min.
19. Store the virus stock in small aliquots at −80°C.
20. Estimate the titer of the virus by adding serial dilutions of virus to $CD34^+$ cells and then plating CFCs in the presence and absence of puromycin as described in **Section 3.1.2**.

3.2.3. Functional Validation Assay for Candidate shRNAs

1. Follow the transduction protocol described in **Section 3.1.2** using 200,000 cells in duplicate for each shRNA vector to be tested including control vector and mock-treated cells.
2. Determine the transduction efficiency 48 h posttransduction by plating CFCs in the presence or absence of puromycin as described in **Section 3.1.2**.
3. Grow cells in long-term culture as described in **Sections 3.1.2** and **3.1.3**, respectively, but perform CFC assays every week starting from week two. Strong phenotypes can be observed as early as two to three weeks by observing increases in overall colony numbers and increasing frequencies of puromycin-resistant colonies relative to controls. Candidates that fail to produce colony numbers or puromycin-resistant colonies in excess of controls after five weeks culture can be considered false-positives.

3.2.4. Target Validation for Candidate shRNAs

1. Candidate shRNAs that have been validated functionally should be tested first for their ability to knockdown their respective target genes.
2. Transduce $CD34^+$ cells as described above with the candidate shRNAs as well as control shRNA using a high MOI (based on step 20 in **Section 3.2.2**) to ensure transduction efficiencies of more than >90%.
3. Harvest the cells after 48 h and extract RNA using RNA purification kit.
4. Complete cDNA synthesis and perform quantitative real-time PCR to determine the expression level for the gene of interest using a housekeeping gene such as HPRT as reference and compare with control-transduced samples.
5. If shRNAs show more than 50% knockdown of their target gene, the specificity of their effect should be further validated by testing additional shRNAs against that gene.

shRNAs generating less than 50% knockdown of their target gene are most likely having 'off-target' effects and should not be analyzed further.

6. Additional shRNAs against a given gene can be obtained from the manufacturer of the library and should be processed and tested for knockdown and function as described for the original shRNA candidates above (**Sections 3.2.2, 3.2.3, and 3.2.4**).
7. If at least two independent shRNAs (including the original candidate) targeting the same gene repeatedly reproduce the phenotype from the screen while simultaneously knocking down target gene transcript levels, the criteria for a positive hit are fulfilled. Subsequent in vitro and in vivo characterizations of the target gene will further clarify its function in stem cell biology.

4. Notes

1. Several different pooled lentiviral shRNA libraries are commercially available from Sigma, Open Biosystems, and System Biosciences. The protocol described here is based on the use of a lentiviral library from The RNAi Consortium (TRC) that is available from Sigma (2). In this library, the pLKO1 self-inactivating lentiviral vector drives shRNA expression from a human U6 promoter and carries the puromycin-resistance gene under control of a PGK promoter.
2. If umbilical cord blood samples are not available, purified $CD34^+$ cells can be purchased from commercial sources.
3. Pooled lentiviral libraries are usually supplied in subpools with around 10,000 shRNAs in each and shipped as ready-to-use viral stocks. If a library is obtained in plasmid DNA format, we refer to **Section 3.2.2** for virus production. To ensure high coverage of the library in the more primitive cell populations (around 5% of the total $CD34^+$ cells), we recommend using at least 3×10^6 $CD34^+$ cells for each subpool of the library to be screened. The cells should be transduced in multiple pools (2×10^5 cells per pool) to allow independent identification of identical shRNAs. Control vectors harboring nonsilencing shRNAs can be obtained from the manufacturer of the library and should be used as reference in the screen (for virus production of control vectors *see* **Section 3.2.2**).

4. Before setting up a screen, we recommend titering the lentiviral library in CD34+ cells since effective titers vary depending on the cell type being transduced. Based on the titer provided by the manufacturer, infect cells at MOIs of 0.5, 1, 2, 5, and 10 and measure transduction efficiency by CFC assays (**Section 3.1.2**). Use an MOI that gives around 35% transduction to avoid doubly-transduced cells in the screen.

References

1. Berns, K., Hijmans, E. M., Mullenders, J., Brummelkamp, T. R., Velds, A., Heimerikx, M., Kerkhoven, R. M., Madiredjo, M., Nijkamp, W., Weigelt, B., Agami, R., Ge, W., Cavet, G., Linsley, P. S., Beijersbergen, R. L., and Bernards, R. (2004) A large-scale RNAi screen in human cells identifies new components of the p53 pathway. *Nature* **428**, 431–437.
2. Moffat, J., Grueneberg, D. A., Yang, X., Kim, S. Y., Kloepfer, A. M., Hinkle, G., Piqani, B., Eisenhaure, T. M., Luo, B., Grenier, J. K., Carpenter, A. E., Foo, S. Y., Stewart, S. A., Stockwell, B. R., Hacohen, N., Hahn, W. C., Lander, E. S., Sabatini, D. M., and Root, D. E. (2006) A lentiviral RNAi library for human and mouse genes applied to an arrayed viral high-content screen. *Cell* **124**, 1283–1298.
3. Paddison, P. J., Silva, J. M., Conklin, D. S., Schlabach, M., Li, M., Aruleba, S., Balija, V., O'Shaughnessy, A., Gnoj, L., Scobie, K., Chang, K., Westbrook, T., Cleary, M., Sachidanandam, R., McCombie, W. R., Elledge, S. J., and Hannon, G. J. (2004) A resource for large-scale RNA-interference-based screens in mammals. *Nature* **428**, 427–431.
4. Westbrook, T. F., Martin, E. S., Schlabach, M. R., Leng, Y., Liang, A. C., Feng, B., Zhao, J. J., Roberts, T. M., Mandel, G., Hannon, G. J., Depinho, R. A., Chin, L., and Elledge, S. J. (2005) A genetic screen for candidate tumor suppressors identifies REST. *Cell* **121**, 837–848.
5. Woods, N. B., Fahlman, C., Mikkola, H., Hamaguchi, I., Olsson, K., Zufferey, R., Jacobsen, S. E., Trono, D., and Karlsson, S. (2000) Lentiviral gene transfer into primary and secondary NOD/SCID repopulating cells. *Blood* **96**, 3725–3733.
6. Ali, N., Karlsson, C., Aspling, M., Hu, G., Hacohen, N., Scadden, D. T., and Larsson, J. (2009) Forward RNAi screens in primary human hematopoietic stem/progenitor cells. *Blood* **113**, 3690–3695.

Chapter 4

A Recessive Genetic Screen for Components of the RNA Interference Pathway in Mouse Embryonic Stem Cells

Melanie I. Trombly and Xiaozhong Wang

Abstract

Several key components of the RNA interference (RNAi) pathway were identified in genetic screens performed in nonmammalian model organisms. To identify components of the mammalian RNAi pathway, we developed a recessive genetic screen in mouse embryonic stem (ES) cells. Recessive genetic screens are feasible in ES cells that are Bloom-syndrome protein (Blm-) deficient. Therefore, we constructed a reporter cell line in Blm-deficient ES cells to isolate RNAi mutants through a simple drug-selection scheme. This chapter describes how we used retroviral gene traps to mutagenize the reporter cell line and select for RNAi mutants. Putative RNAi mutants were confirmed using a separate functional assay. The location of the gene trap was then identified using molecular techniques such as Splinkerette PCR. Our screening strategy successfully isolated several mutant clones of Argonaute2, a vital component of the RNAi pathway.

Key words: Bloom-deficient mouse embryonic stem cells, RNA interference, recessive genetic screen, retroviral gene-trap mutagenesis, splinkerette PCR.

1. Introduction

One obstacle to performing a recessive genetic screen in mammalian cells is the diploid nature of the genome. A recessive mutation must be rendered homozygous in order to observe a mutant phenotype. This challenge has been partially overcome by the use of Bloom's syndrome protein (Blm)-deficient mouse embryonic stem (ES) cells (1, 2). The Bloom gene encodes a homologue of the bacterial RecQ helicase and loss of this gene leads to genome instability (3). An 18- to 27-fold higher loss-of-heterozygosity (LOH) has been reported for Blm-deficient ES

B. Zhang, E.J. Stellwag (eds.), *RNAi and MicroRNA-Mediated Gene Regulation in Stem Cells*, Methods in Molecular Biology 650, DOI 10.1007/978-1-60761-769-3_4,

cells compared to wild-type ES cells (2, 3). Therefore, these cells convert heterozygous mutations to homozygous mutations at a much higher frequency. Several pioneering studies have shown that Blm-deficient ES cells are amenable to forward phenotypic screens (1, 2, 4).

A genetic screen performed in Blm-deficient ES cells depends on a powerful selection scheme to distinguish mutant cells from the rest of the population. To establish a selection system for RNAi mutants in Blm-deficient ES cells, we chose Hypoxanthine phosphoribosyl transferase (*Hprt*) as the reporter gene because it allows for both positive and negative selection. When stably expressed in the cells, *Hprt* confers resistance to the drug hypoxanthine aminopterin thymidine (HAT^R) and sensitivity to the drug 6-thioguanine ($6\text{-}TG^S$) (**Fig. 4.1A**). By introducing a small-hairpin RNA (shRNA) against the *Hprt* gene, we can select for cells in which the reporter gene is silenced (cells become 6-TG

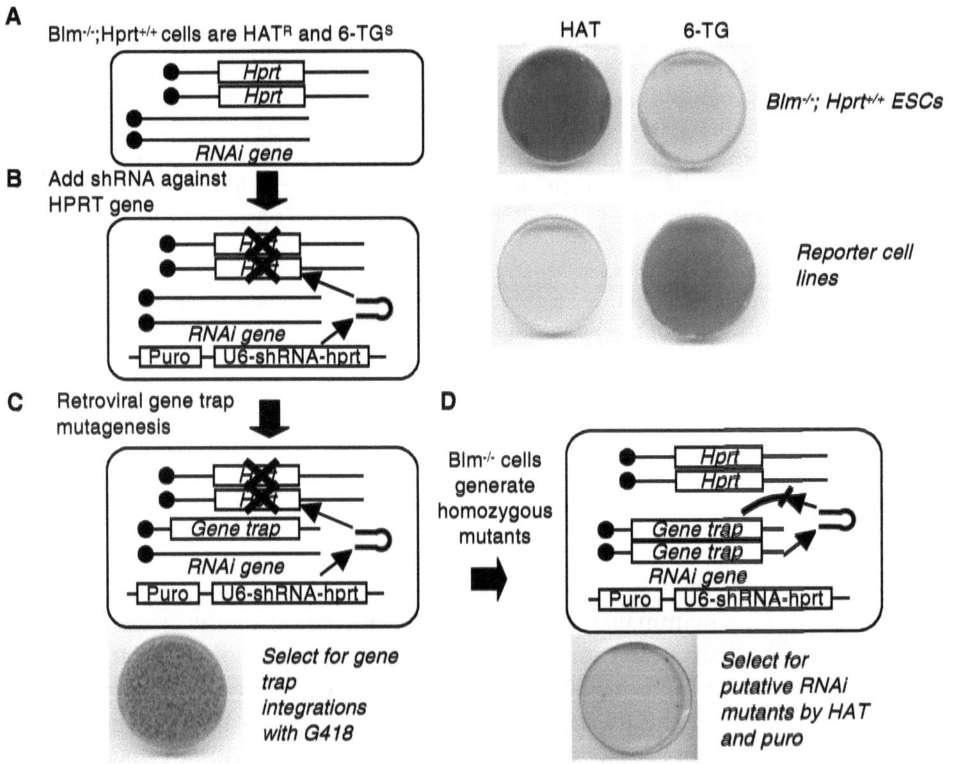

Fig. 4.1. Developing a recessive genetic screen in Blm-deficient ES cells for RNAi mutants. (**A**) The parental cell lines are Blm-deficient and contain two copies of a targeted PGK-*Hprt* mini gene. The *Hprt* gene confers HAT^R and $6\text{-}TG^S$ to the cells. Shown on the right is a methylene blue stain for cells. (**B**) Introducing a puromycin-linked shRNA against the *Hprt* gene causes the cells to become HAT^S and $6\text{-}TG^R$ through a working RNAi pathway. (**C**) The reporter cell lines were then mutagenized with a retroviral gene trap and gene-trap integrations were selected with G418. Methylene blue staining of $G418^R$ colonies is shown below. (**D**) After passaging to create homozygous mutants for the gene trap, the cells were selected with HAT and puromycin to isolate putative RNAi mutants. Methylene blue staining of HAT^R, $puro^R$ colonies is shown below. (Modified from (5) with permission from Oxford University Press.)

resistant, 6-TG^R) through a working RNAi pathway (**Fig. 4.1B**). More importantly, after mutagenesis, the reporter gene allows for positive selection of cells that become RNAi-deficient.

We constructed our selection system in a Blm-deficient ES cell line containing two copies of the PGK-*Hprt* minigene sequentially targeted at the mouse *gdf7* locus (1) (**Fig. 4.1A**). Having two copies of the *Hprt* gene insures a homozygous state required in a loss of heterozygosity (LOH)-based recessive genetic screen. We electroporated *Hprt*-expressing cells with a U6-promoter-driven shRNA to silence the *Hprt* gene. The shRNA contained a puromycin (puro) marker (puro::shRNA) to select for cells that stably incorporated the transgene. Blm-deficient cells expressing the puro::shRNA effectively silenced the *Hprt* gene and the cells became puromycin resistant ($puro^R$), HAT sensitive (HAT^S), and 6-TG^R (**Fig. 4.1B**). We established several $puro^R$, HAT^S, and 6-TG^R stable cells lines for screening. The reporter cell lines were mutagenized with a retroviral gene trap and then passaged to create homozygous mutants for the gene trap (**Fig. 4.1C**). Finally, cells were selected with HAT and puromycin to isolate putative RNAi mutants (**Fig. 4.1D**).

Using retroviral gene traps, we successfully performed a genome-wide screen for RNAi mutants in our reporter cell lines (5). Our screen led to the isolation of several clones that were homozygous for gene-trap integrations in intron1 of Argonaute2 (Ago2). Thus, the identification of Ago2, a known essential component of RNAi, validates the utility of our screening strategy. With future improvements in insertional mutagenesis and an increase of genomic coverage, our system has the potential to identify other unknown components in the mammalian RNAi pathway.

2. Materials

2.1. Cell Culture of Reporter ES Cell Lines and SNLi/SNLPi

1. 10-cm tissue culture grade plates.
2. 15-mL conical tubes.
3. M15 medium for ES cells: Knockout Dulbecco's Modified Eagle's Medium (Knockout-DMEM) supplemented with 15% fetal bovine serum (FBS, Hyclone ES-cell grade), 1× of penicillin–streptomycin–glutamine (100 × PSG, Invitrogen), and 100 μM of betamercaptoethanol (β-ME, Invitrogen). Stored at 4°C for up to 1 month.
4. M10 medium for feeder cells: Knockout-DMEM supplemented with 10% fetal bovine serum (FBS, Atlanta Biologicals) and 1× of penicillin–streptomycin–glutamine

(100×) liquid (Invitrogen). Stored at 4°C for up to 1 month.

5. Trypsin (0.25% trypsin and ethylenediamine tetraacetic acid, Invitrogen).
6. Phosphate-buffered saline (PBS).
7. 0.1% gelatin (autoclaved 0.1% gelatin from porcine (Sigma) prepared with ddH_2O, stored at room temperature).
8. Reporter ES cell line that is $Blm^{-/-}$, homozygous for the *Hprt* reporter gene and contains an active shRNA against *Hprt*.
9. Mitomycin-treated feeder cells that are neomycin and neomycin/puromycin resistant (SNLi and SNLPi, respectively) (*see* **Note 1**).
10. Tissue culture incubators (37°C and 5% CO_2).

2.2. Calcium Phosphate Transfection of Phoenix Cells for Virus Production

1. 50-mL conical tubes.
2. 10-cm tissue culture grade plates.
3. 0.1% gelatin.
4. Phoenix cells.
5. M10 media for Phoenix cells (stored at 4°C for up to 1 month): Knockout-DMEM, 10% heat-inactivated (55°C for 30 min), fetal bovine serum (FBS, Atlanta Biologicals), and 1× PSG.
6. Trypsin.
7. PBS.
8. Calcium phosphate transfection kit (Clontech).
9. Bleach for treating materials that come in contact with virus.
10. DNA plasmids for retroviral vectors: PGGV2, PGGV5, PGGV6, or PGGV7 prepared by Qiagen Midi Kit and purified by phenol/chloroform extraction.

2.3. Infection of ES Cells and Selection for Putative RNAi Mutants

1. 96-well flat-bottom and U-bottom tissue culture grade plates and 10-cm tissue culture grade plates.
2. 0.1% gelatin.
3. 12-channel pipettor.
4. 8-channel aspirator.
5. 0.45-μm filter system for filtering virus.
6. Polybrene at 1,000× (4 mM) (hexadimethrine bromide, Sigma), store at 4°C.
7. SNLi and SNLPi feeders.
8. 50× geneticin (G418, Invitrogen), 50× hypoxanthine aminopterin thymidine (HAT, Invitrogen), 10 μM

6-thioguanine (6-TG, Sigma), and 2 mg/mL puromycin (Invivogen).

9. PBS.

10. Leukemia inhibitory factor, LIF (Esgro, 10^7 U/mL, Millipore), used as 10,000× stock.

2.4. Dual-Luciferase Assay to Confirm RNAi Mutant Phenotype

1. Dual-luciferase kit (Promega).
2. 24-well and 6-cm tissue culture grade plates.
3. 0.1% gelatin.
4. Opti-MEM (Invitrogen).
5. Lipofectamine 2000 (Invitrogen).
6. DNA plasmids (for Renilla luciferase, Firefly luciferase, shRNA against firefly luciferase, and control shRNA) prepared by Qiagen Midi Kit and purified by phenol/chloroform extraction.
7. $20/20^n$ single-tube luminometer (Turner BioSystems).

2.5. Splinkerette PCR Analysis of Gene-trap Integration Site

1. 6-cm tissue culture grade plates.
2. 0.1% gelatin.
3. Capillary tubes 1.5 × 90 mm.
4. Lysis buffer: 50 mM Tris pH 7.5, 10 mM EDTA pH 8.0, 100 mM NaCl, 0.5% SDS, and 1 mg/mL Proteinase K (added fresh).
5. NaCl/EtOH mixture: 0.075 M NaCl, 100% EtOH mixture for precipitating DNA (prepared fresh).
6. Primer sequences of oligos required:

   For PGGV2 (AB949): 5′-GCT AGC TTG CCA AAC CTA CAG GTG G-3′

   For PGGV2 (HM001): 5′-GCC AAA CCT ACA GGT GGG GTC TTT-3′

   For PGGV2-V7 (HMSP1): 5′-CGA AGA GTA ACC GTT GCT AGG AGA GAC C-3′

   For PGGV2-V7 (HMSP2): 5′-GTG GCT GAA TGA GAC TGG TGT CGA C-3′

   For PGGV5-7 (1748): 5′-TAG GTC ACT CGA CCT GCA GAC C-3′

   For PGGV5-7 (1749): 5′-TCG ACC TGC AGA CCA AGA TCG CT-3′

   Splinkerette linkers for Sau3A1 digest, HMSpAa: 5′-CGA AGA GTA ACC GTT GCT AGG AGA GAC CGT GGC TGA ATG AGA CTG GTG TCG ACA CTA GTG G-3′

Splinkerette linkers for Sau3A1 digest, HMSpBb: 5′-GAT CCC ACT AGT GTC GAC ACC AGT CTC TAA TTT TTT TTT TCA AAA AAA-3′

7. Sau3A1 enzyme and buffer, T4 DNA ligase, and NEB Buffer 2 (New England Biolabs).
8. Reagents for PCR: Advantage cDNA polymerase (Clontech), 10× cDNA buffer, 5 mM dNTPs, molecular biology H_2O.
9. Thermal cycler.
10. Qiagen PCR purification kit and gel purification kit.
11. TOPO TA cloning kit for sequencing (Invitrogen).

3. Methods

Retroviral gene-trap mutagenesis has been successfully used in recessive genetic screens performed in Blm-deficient ES cells (1, 4). A major advantage of this approach is the ability to use the gene-trap provirus as a molecular tag to identify the gene-trap integration site. Integration within a gene results in splicing of upstream exons to the gene trap, and a polyadenylation signal in the gene trap prematurely terminates the mRNA leading to loss of downstream exons (**Fig. 4.2B**). The gene trap also contains a neomycin-resistance marker to allow for geneticin (G418) selection of the stably integrated and expressed gene traps. We performed a genome-wide screen for RNAi components in our reporter cell lines ($Blm^{-/-}$ $Hprt^{+/+}$ ES cells expressing shRNA against *Hprt*) using a recombinant retroviral gene trap.

To increase the genomic coverage of the screen, four different retroviral gene-trap constructs were used (**Fig. 4.2A**). The PGGV2 retroviral gene-trap vector was used in previous recessive screens employing Blm-deficient ES cells (1, 4). Several modifications to PGGV2 were incorporated to create the following retroviral gene-trap vectors: PGGV5, PGGV6, and PGGV7 (**Fig. 4.2A**). PGGV5 contained a unique sequence inserted into the LTR of PGGV2 to facilitate subsequent molecular cloning of the integration. PGGV6 was a modified version of PGGV5 containing an additional promoter to drive neomycin resistance independent of the gene-trap integration site. PGGV7 was a modified version of PGGV6 with an additional splice-acceptor site.

The reporter ES cell lines were infected with gene-trap retroviruses produced from a Phoenix packaging cell line. ES cells containing retroviral integrations that were active gene traps were selected by G418 resistance (**Fig. 4.1C**). After distinct $G418^R$

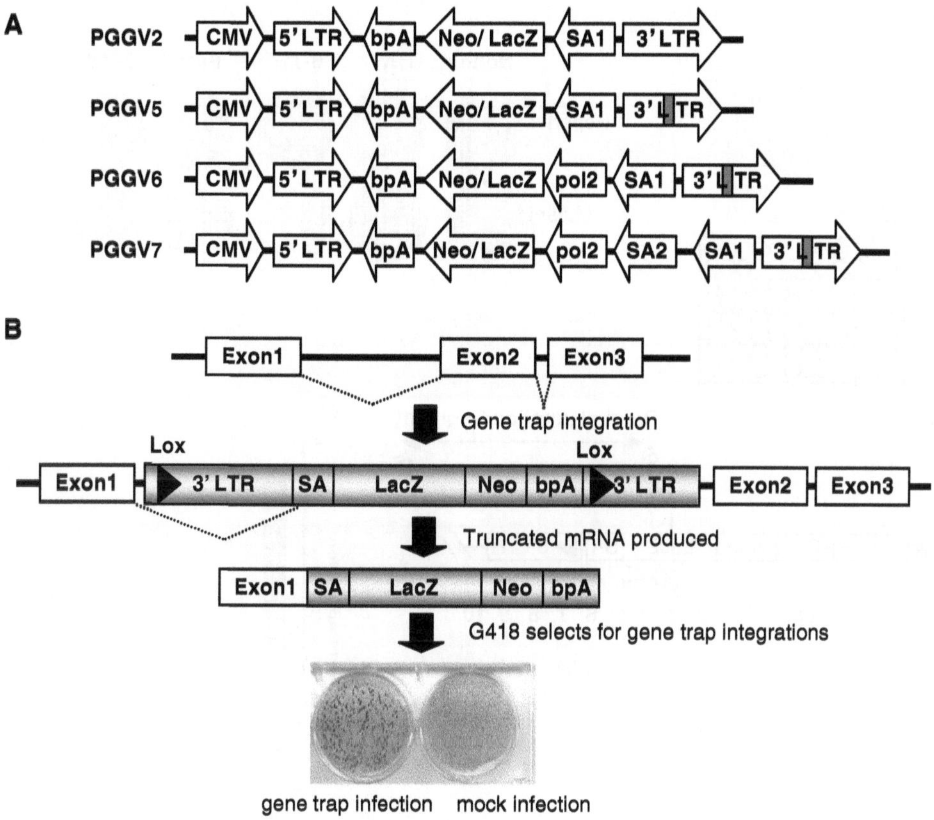

Fig. 4.2. Diagram of retroviral gene-trap mutagenesis. (**A**) A diagram showing the four retroviral constructs used. (**B**) An example of how gene-trap integration can create mutations. The gene trap used contains a Neomycin marker so that gene-trap integration events can be selected for with G418. Abbreviations used: CMV = cytomegalovirus, LTR = long terminal repeat, bpA = polyadenylation, Neo = Neomycin resistance marker, LacZ = beta-galactosidase, pol2 = PolII promoter, SA = splice acceptor site. (Modified from (5) with permission from Oxford University Press.)

colonies had formed, the cells were propagated twice to allow accumulation of homozygous recessive mutations for the gene trap. Putative RNAi mutant cells were selected with HAT for their ability to regain *Hprt* expression. HAT^R colonies were also selected with puromycin to ensure that the puro::shRNA was retained (**Fig. 4.1D**).

After isolation of putative mutants, several assays can be used to eliminate false positives and facilitate analysis of unique mutants. First, to verify the drug-resistance profile, colonies are transferred to 96-well plates and replica plated for selection by HAT, puro, and G418 (**Fig. 4.3A**). RNA, DNA, and protein samples from mutant clones that pass all drug-resistance tests are collected for further analyses. Second, to avoid redundant analysis of identical mutant clones, a Southern blot can be performed on the clones. Southern blot analyses will reveal which clones isolated from the same pool contain a similar proviral–host genome junction fragment based on size. This occurs from

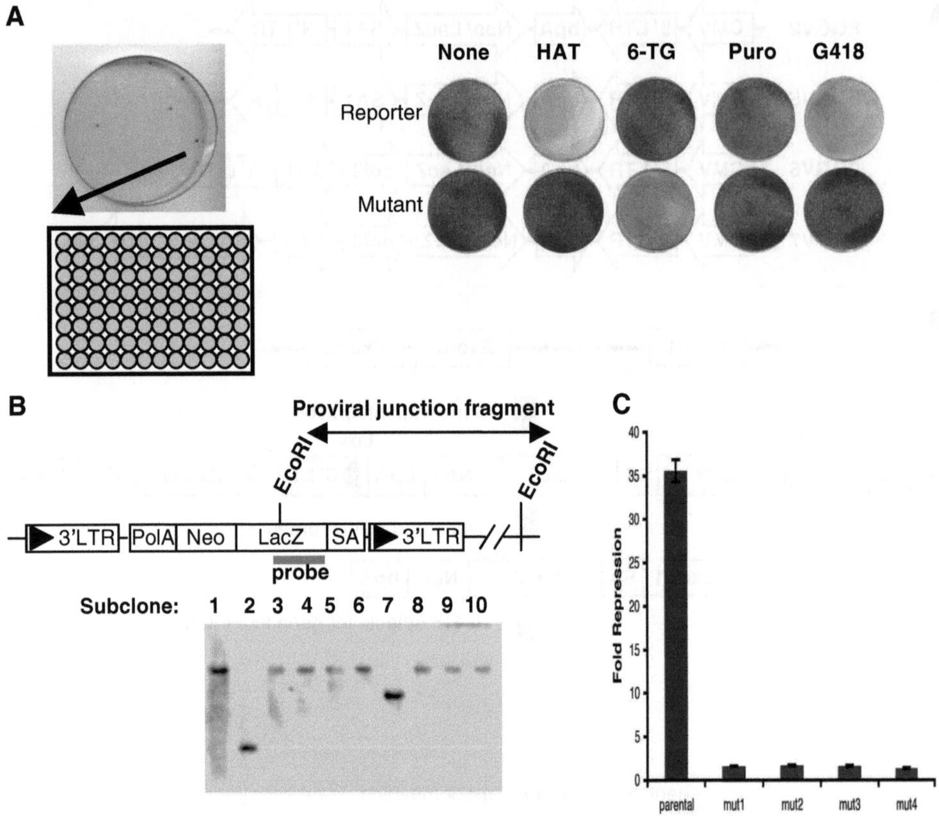

Fig. 4.3. Further analyses and validation of mutants. (**A**) Mutants isolated by HAT and puro selection are picked to 96-well plates where they are expanded for stock, confirmation of drug resistance, DNA, and RNA. To the right is a methylene blue stain of cells under various drug selections to confirm the mutant phenotype. (**B**) Southern blot of DNA prepared from clones isolated from the same gene-trap pool reveals that the majority of clones have the same gene-trap integration. (**C**) A luciferase assay to confirm that the mutants are unable to silence a reporter gene. (Modified from (5) with permission from Oxford University Press.)

daughter clones that arise from passaging the cells (**Fig. 4.3B**). In addition, Southern blot analyses should reveal a single band for the majority of the clones, which verifies that the retroviral gene trap did not integrate more than once in the genome. Finally, to eliminate false positives, a secondary functional assay should be performed to confirm the RNAi mutant phenotype of the isolated clones. A Firefly luciferase (F-luc) reporter is transfected into the ES cells to measure the repression of an shRNA against F-luc. Renilla luciferase (R-luc) serves as an internal control for transfection efficiency. The ability of the shRNA to repress F-luc is assessed in both mutant cell lines and reporter cells lines and mutants that have lost the ability to repress the reporter are used for further characterization (**Fig. 4.3C**).

After confirmation of the RNAi mutant phenotype by luciferase assay, the next step is to determine the genomic location of the gene trap. Several different molecular methods can be

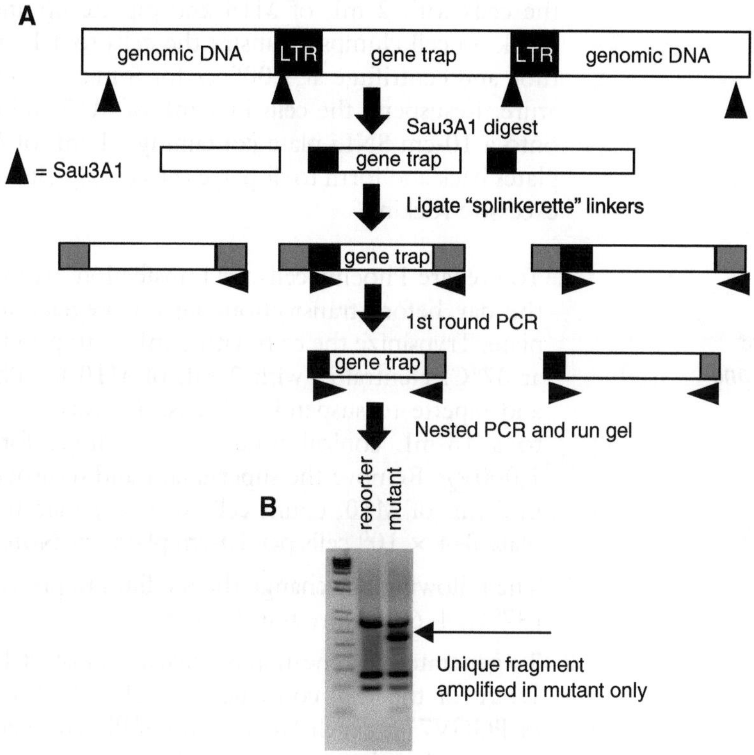

Fig. 4.4. Splinkerette PCR to identify gene-trap integration site. (**A**) Genomic DNA is isolated from mutant and reporter cell lines. Sau3A1 digest is used to fragment the DNA. Splinkerette linkers are then ligated to the fragments and two rounds of PCR are used to amplify the genomic sequence flanking the gene-trap integration. (**B**) The gel shows an example of an ideal Splinkerette reaction electrophoresed on a gel where a unique band, amplified only in the mutant cell line, is indicated by the *arrow*.

used to identify the genomic location; however, we found that Splinkerette PCR was the most effective way to determine the integration site for the majority of our clones (**Fig. 4.4**). Splinkerette PCR is a modified form of PCR walking in which specialized linkers are ligated to the DNA fragments to enhance specific amplification of the desired product (6).

3.1. Culturing of Reporter ES Cells

1. Plate 10 cm of SNLi or SNLPi cells 2 days before ES cells will be plated on top of them. SNLi/SNLPi cells need time to adhere and spread out to create a good cell surface layer for ES cells to grow on.
2. The reporter ES cells should be grown on 10-cm plates of SNLi or SNLPi feeder cells and passaged at 1:4–1:6 every 2 days (*see* **Note 2**).
3. To passage ES cells, feed them at least 2 h before passaging with M15 medium. Aspirate off the media and wash the 10-cm plate once with 10 mL of PBS. Trypsinize the cells with 2 mL of trypsin at 37°C for 15 min. Neutralize

the cells with 2 mL of M15 and pipette up and down to break up cell clumps. Transfer the cells to a 15-mL conical tube and centrifuge at 1,000×*g* for 5 min at room temperature. Resuspend the cells in 4 mL of M15 and plate 1 mL onto a 10-cm SNLi plate containing 11 mL of M15. Rock plates back and forth to disperse cells evenly. Be sure to avoid excessive rocking.

3.2. Calcium Phosphate Transfection of Phoenix Cells and Virus Harvest

1. To prepare Phoenix cells for transfection, passage the cells the day before transfection onto a pregelatinized 10-cm plate. Trypsinize the cells with 2 mL of trypsin for 2–5 min at 37°C. Neutralize with 2 mL of M10 for Phoenix cells and pipette to suspend the cells. Transfer cell suspension to a 15-mL conical tube and centrifuge for 5 min at 1,000×*g*. Remove the supernatant and resuspend the cells in 5 mL of M10, count cells with a hemacytometer, and plate 3–4 × 10^6 cells per 10-cm plate (*see* **Note 3**).
2. The following day change the medium to prewarmed M10 (37°C) 4–6 h before transfection.
3. To generate the gene-trap retrovirus, transfect 12 μg of the DNA for the viral construct (PGGV2, PGGV5, PGGV6, or PGGV7) for each 10-cm plate of Phoenix cells using the calcium phosphate transfection kit (following the manufacturer's instructions).
4. Before harvesting virus, it is important to remove the transfection reagent by washing with PBS. Approximately 16 h after transfection, wash the transfected Phoenix cells 3 times with 10 mL of PBS. Be careful not to disturb the cells as they detach easily (*see* **Note 4**).
5. To begin collection of virus-containing medium, add prewarmed M10 medium (10 mL) per plate.
6. After 8 h, collect the virus-containing medium by carefully pipetting off the medium into a 50-mL conical tube. Replenish by gently adding 14 mL of prewarmed M10 to the sidewall of the plate. Incubate for 16 h. *Pipettes and any containers that have been used for virus must be treated with bleach before discarding.*
7. Harvest the virus-containing medium in the morning and replenish with 10 mL of prewarmed M10 gently to avoid cell detachment.
8. Harvest the virus-containing medium in the evening and replenish with 14 mL of prewarmed M10 gently to avoid cell detachment.
9. Harvest the final virus-containing medium in the morning and bleach the 10-cm Phoenix plates before discarding.

10. To remove any dead cells that might have detached into the virus-containing medium, combine all virus harvests and filter through a 0.45-μm filter. Prepare supplemented virus medium by adding Hyclone/FBS serum to bring the total serum concentration to 15% (i.e., add 5% more Hyclone/FBS) and adding 100 μM β-ME.

3.3. Infection of ES Cells and Isolation of Mutants

1. To begin preparing the ES cells for infection, plate 15 × 10-cm plates of SNLi cells 3 days prior to infection.
2. The day before infection, plate the reporter ES cell lines at 3×10^6 cells per 10 cm onto the SNLi (*see* **Note 5**).
3. The following day, infect cells by adding 10 mL of the supplemented virus to the plate and adding 1× polybrene to precipitate the virus and help increase the titer.
4. At night, add 5 mL more of supplemented virus and 1× polybrene.
5. The following morning, remove the supplemented virus by aspirating it into a container with bleach. Then replenish the media on the ES cells by adding 10 mL of M15 medium per 10-cm plate.
6. To isolate the ES cells containing an actively expressed retroviral gene-trap integration event, start selection with geneticin (G418). In the afternoon, add G418 directly to the plate to obtain a final concentration of 300 μg/mL (e.g., 60 μL of 50 mg/mL G418 into 10 mL of M15); G418 will change the pH of the medium and turn it an orange-yellow color when added. Rock the plates gently to insure even distribution of the G418 into the medium.
7. The following morning there should be many dead floating cells in the medium; remove the medium and replace it with fresh 350 μg/mL G418 in M15.
8. Continue to change medium every 1–2 days with fresh 350 μg/mL G418 in M15.
9. Approximately 5 days after beginning G418 selection, prepare 10-cm SNLi plates (1 for every 10-cm plate of infected cells) to be able to passage the $G418^R$ colonies.
10. After about 1 week in selection, distinct $G418^R$ colonies should form. The cells then need to be passaged to allow for mitotic recombination to occur and create cells that are homozygous for the gene trap. Wash the infected plates once with 10 mL of PBS. Trypsinize with 2 mL of trypsin for 17 min to make sure that the colonies are sufficiently broken up into single cells. Neutralize with 2 mL of M15 medium and pipette up and down several times to disperse the cells. Place them into a 15-mL conical tube and centrifuge for 5 min at 1,000×*g*. Remove the supernatant

and resuspend the pellet in 1 mL of M15 to plate onto a 10-cm plate containing 11 mL of M15.

11. One day after passaging, the cells will be confluent and need to be passaged onto SNLPi plates for the next selection step. Wash and trypsinize as in Step 10 and plate approximately 10×10^6 of the infected G418^R cells on each 10-cm plate in M15 medium containing 0.5 U/mL LIF (2 × 10 cm plates per infected 10-cm plate, approximately $^1/_4$ of each).
12. To isolate the G418^R cells that are putative RNAi mutants, begin selection for cells that no longer repress the *Hprt* gene. The following morning add 1× HAT + 1U/mL LIF in M15 to the plates (10–12 mL per plate).
13. The next day change to 1× HAT + 1U/mL LIF in M15, 10–12 mL per plate. There should be many dead floating cells and the medium will be orange colored.
14. The next day change to 1× HAT + 1U/mL LIF in M15, 10–12 mL per plate. There should be fewer dead floating cells.
15. The media should need minimal changing for the next 5 days, changing every 2–3 days as needed. Look at the plates daily to check for colony formation.
16. After the HAT^R colonies have formed, it is important to confirm that the colonies contain the puro::shRNA and release them from HAT selection. A week after starting HAT selection, change medium to 1× HT + 2 μg/mL puromycin + 1U/mL LIF in M15.
17. Prepare 96-well SNLPi plates to prepare for colony transfer and selection with individual drugs HAT, 6-TG, and puro.
18. Three to five days after starting puromycin selection and releasing the cells from HT, putative RNAi mutant colonies will have formed. Pick HAT^R/Puro^R colonies to expand for future analyses. For picking colonies, prepare a U-bottom 96-well plate containing 40 μL/well of trypsin. Wash the 10-cm plate containing colonies with 10 mL of PBS to remove medium. Remove PBS from the plate and replace with 5 mL of PBS. Pick colonies under a laminar biosafety cabinet using a dissection microscope. Colonies can be mechanically dislodged by drawing a circle around the colony with a 200-μL pipette tip and then sucking up the colony into the pipette tip (200-μL pipettor set at 15 μL). Alternatively, colonies can be picked by aspirating the colony while simultaneously poking at the colony with a 200-μL pipettor set at 15 μL. Place a colony in the U-bottom well containing trypsin and continue to the next colony (*see* **Note 6**). When all colonies have been

picked, incubate the 96-well U-bottom plate at 37°C for 17 min. Using a 12-channel pipettor, neutralize each well with 40 μL of 1U/mL LIF in M15. Pipette each well up and down 20 times to disperse colonies into individual cells. Plate entire cell suspension into a 96-well flat bottom SNLi plate containing 150 μL of 1U/mL LIF in M15.

19. Change the medium the following morning (making sure cells have attached first) to fresh 200 μL of 1U/mL LIF in M15.
20. After 3–4 days, the 96-well plate should be ready to passage. Passage cells 1:4 onto 4 × 96-well plates: 1 plate for stock, 1 plate for HAT selection, 1 plate for puromycin selection, and 1 plate for G418 selection. Using a 12-channel pipettor and 8-channel aspirator, wash the 96-well plate 1× with PBS and trypsinize with 40 μL of trypsin for 17 min at 37°C. Neutralize with 90 μL of M15 containing 1U/mL LIF. Pipette each well up and down 20 times to disperse colonies into individual cells. Plate 40 μL of each well into 96-well plates containing 150 μL of M15 with 1U/mL LIF and the appropriate drug (or no drug for the stock plate). After 2–3 days in individual drug selection, make a note of which cells are resistant to all three drugs: HAT, puro, and G418. These cells are putative mutants. The rest do not pass all drug tests and should not be analyzed further.
21. Expand putative mutants to 6 wells and freeze stocks. If several mutants are isolated from the same pool a Southern blot should be performed to identify unique clones.

3.4. Dual-Luciferase Assay to Confirm Mutants

The putative RNAi mutants can be confirmed using a dual-luciferase assay. In this assay, firefly luciferase is used as a reporter of shRNA-directed cleavage and renilla luciferase serves as the internal control for the transfection efficiency. The ability of the putative mutants to repress the firefly reporter is compared to the repression of the original reporter cell line before mutagenesis.

1. To prepare cells for transfection of the luciferase constructs, it is important to remove the feeder cells as they can also be transfected. This can be achieved by passaging the ES cells on pregelatinized plates. Passage mutant cells as well as the original nonmutagenized reporter cells (as a control) onto 6-cm pregelatinized plates in M15 with 1U/mL LIF for two consecutive passages.
2. Plate both mutant and reporter cells onto a pregelatinized 24-well plate at 150,000 cells/well in M15 + 1U/mL LIF. Plate 6 wells of each mutant and 6 wells of the parental cells in order to do triplicates for each transfection.

3. The following day change the medium to 0.5 mL Opti-MEM, 2 h before transfection.
4. Follow the Invitrogen Lipofectamine 2000 directions to transfect DNA. CMV-Renilla-luciferase acts as an internal control and CMV-Firefly-luciferase as a reporter. One set will be transfected with control U6-shRNA; the other set will be transfected with U6-shRNA against firefly luciferase. Prepare a master mix of each transfection with the following DNA concentrations:

DNA	Concentration per well (ng)
Transfection 1	
CMV-F-luc	100
CMV-R-luc	20
U6-shRNA control	250
Transfection 2	
CMV-F-luc	100
CMV-R-luc	20
U6-shRNA luc1	250

5. Change medium to M15 + 1U/mL LIF the following morning.
6. 48 h after transfection, harvest cell lysates for assay using the Promega Dual luciferase kit and take readings on the $20/20^{n}$ luminometer using clear microcentrifuge tubes.

3.5. Splinkerette PCR for Gene-trap Integration Analysis

After confirmation of the RNAi mutant phenotype by luciferase assay, the next step is to analyze the gene-trap integration site using Splinkerette PCR, a modified form of PCR walking. Due to the presence of nonspecific bands that are sometimes amplified during Splinkerette PCR, it is important to conduct a side-by-side control reaction with nonmutagenized reporter cells. First, DNA is obtained from both the mutant and the nonmutagenized reporter cell line. The DNA is then digested with a frequent four base pair cutter, in this instance Sau3A1. The fragments are then ligated to the Splinkerette linkers. The ligated products are used for two rounds of PCR using primer sets that anneal to a unique region in the LTR of the gene trap and the Splinkerette linker. The PCR products from the nonmutant and the mutants are compared to find a unique band only amplified in the mutant that represents the genomic region flanking the gene trap.

1. To analyze the gene-trap integration by Splinkerette PCR, first obtain DNA from the mutant and nonmutant cells.

Plate 3 × 10^6cells in a pregelatinized 6-cm plate in MI5 with 1U/mL LIF. Prepare one plate each of mutant and nonmutant cells. Grow until cells are confluent, changing the medium daily.

2. To obtain DNA from the cells, wash them 1× with PBS. Then add 1 mL of lysis buffer to each 6-cm plate. Using a transfer pipette, pipette the lysis buffer in the plate to lyse the cells. Then transfer the lysate to a 15-mL conical tube. The lysate will be extremely viscous. Place in 37°C incubator to lyse overnight.
3. To perform DNA extraction, first remove conical tube from incubator and cool to room temperature (RT). To precipitate the DNA, add an equal volume of the 0.075 M NaCl/100% EtOH mixture to the lysate (1 mL). Invert the tube several times till a white fluffy precipitate forms. Remove the precipitate with a glass rod and wash the DNA briefly by swirling the glass rod with the DNA in a 1.5-mL microcentrifuge tube containing 1 mL of 70% EtOH. The DNA should remain attached to the glass rod during this step. To resuspend the DNA, transfer it from the glass rod into a 1.5-mL microcentrifuge tube containing 0.5-mL molecular biology grade H_2O. Swirl the glass rod until the DNA falls off. Vortex the DNA and H_2O mixture and then heat for 30 min to 1 h at 55°C to allow the DNA to solubilize. Repeat the precipitation step and resuspend the DNA in 400 μL of molecular biology H_2O; then store at −20°C until needed.
4. To create the Splinkerette linkers, resuspend the HMSpAa and HMSpBb oligos (HPLC purified) to 50 pmol/μL stock. Then anneal them in a mixture containing NEB Buffer 2 used as a 20× annealing buffer as follows:

Reagent	Volume (μL)
50 pmol/μL HMSpAa	3
50 pmol/μL HMSpBb	3
10× NEB buffer 2	5
ddH_2O	89
Total:	100

Bring the oligo mixture to 65°C for 5 min and cool to RT gradually. The annealed Splinkerette oligos can be used after cooling or stored at −20°C. Incubate the Splinkerette linkers at 65°C for 5 min and cool to RT before each use.

5. To create compatible ends for ligating to the Splinkerette linkers, digest 5 μg of DNA (for both the control and mutant) with 2.5 μL of Sau3A1 in a final volume of 40 μL at 37°C for 3 h, followed by 65°C for 20 min to inactivate the enzyme.
6. Set up a ligation reaction to ligate the Splinkerette linkers to the Sau3A1-digested genomic DNA as follows:

Reagent	Volume (μL)
Sau3AI-digested DNA	6.7
Splinkerette linkers	3
T4 DNA ligase (400 U/μL)	1
10× ligation buffer	2
ddH_2O	7.3
Total:	20

Ligate at 15°C overnight and the following morning replenish with 0.5 μL ligase and return to 15°C for 4 more h.

7. The ligated DNA should be purified to remove unligated Splinkerette linkers by drawing it through a column such as the Qiagen PCR purification Kit and eluting into 30 μL with elution buffer.
8. The purified ligation mixture is ready for the first round of PCR with primers that anneal to the Splinkerette linker and a unique site in the gene trap (*see* **Fig. 4.4**). The first round of PCR should be set up as follows[a]:

Reagent	Volume (μL)
Purified DNA	25
AB949 primer (10 μM)	2
HMSpl primer (10 μM)	2
10× PCR buffer	5
5 mM dNTPs	5
Advantage cDNA polymerase	0.7
ddH_2O	10.3
Total:	50

[a]For Splinkerette when PGGV2 is used. When PGGV5-V7 are used replace AB949 with 1748 primer.

PCR condition:

Denature:	94°C, 1:30
2 cycles:	94°C, 1:00
	64°C, 0:30
	68°C, 2:00
30 cycles:	94°C, 0:30
	64°C, 0:30
	68°C, 2:00
Final extension:	68°C 10:00
Hold:	4°C

After the first round of PCR, the PCR products need to be diluted 1:500 using molecular biology water before proceeding to the next round of PCR; this reduces the amount of the first-round PCR primers left in the mixture. Set up the Nested PCR as follows[a]:

Reagent	Volume (μL)
Diluted 1st round PCR product	25
HMSp2 primer (10 μM)	2
HM001 primer (10 μM)	2
10× PCR buffer	5
5 mM dNTPs	5
Advantage cDNA polymerase	0.7
ddH_2O	30.3
Total:	50

[a]For Splinkerette when PGGV2 is used. When PGGV5-V7 are used, replace HM0001 primer with 1749 primer.

Nested 2 PCR condition:

Denature:	94°C, 1:30
30 cycles:	94°C, 0.30
	61°C, 0:30
	68°C, 2:00
Final extension:	68°C, 10:00
Hold:	4°C

Analyze the PCR reaction products from the mutant and non-mutant DNA side-by-side by electrophoresis on a 2% agarose gel. Compare the PCR products from the mutant cells to the reporter cells to identify the unique band(s) amplified only from the mutant DNA (*see* **Fig. 4.4** for example). Gel purify the unique band using the Qiagen gel purification kit and TOPO clone the band using the TOPO TA Cloning Kit for Sequencing. Miniprep resulting clones and sequence using T7 and T3 primers to identify the flanking genomic region prepared from the mutant.

4. Notes

1. For best results, feeders should be plated at least 2 days before ES cells will be plated. Feeders are generally good for up to 2 weeks after plating.
2. ES cell media should never be allowed to turn yellow as this is unhealthy for the cells, so frequent feeding (once a day) is often necessary.
3. While culturing Phoenix cells, avoid letting the cells become overly confluent as this will reduce the titer of the virus. Each 10-cm plate yields about 48 mL of virus, so plan to produce enough virus-containing medium to be able to infect each 10-cm plate of ES cells with 15 mL of virus.
4. Using a large orifice pipette, such as a 25-mL pipette, and slowly adding PBS to the sidewall of the dish can help prevent cells from detaching.
5. When culturing cells for infection, make sure to have enough ES cells to infect (one confluent 10-cm plate is usually good for plating up to 6 × 10-cm plates for infection). For example, to be able to infect 15 × 10-cm plates, you should have 2–3 × 10-cm plates of ES cells growing in order to have enough cells. To calculate the titer of the virus, you can also plate one additional 10-cm or 6-cm plate, which can be used for methylene-blue staining of $G418^R$ colonies.
6. Try to group colonies isolated from the same infection pool if a Southern blot will be performed to identify identical daughter clones.

Acknowledgments

The authors would like to thank A. Bradley and G. Guo for providing PGG5-4 ES cells and the PGGV2 gene-trap construct and G. Guo and W. Wang for technical advice. H. Su and D. Trombly provided critical reading and comments on the chapter. Funding was provided by National Institute of Health (5R21GM079528); Illinois Department of Public Health (to X.W.); and CMBD training grant from the National Institute of Health T32 GM008061 (to M.T., partial).

References

1. Guo, G., Wang, W., and Bradley, A. (2004) Mismatch repair genes identified using genetic screens in Blm-deficient embryonic stem cells. *Nature* **429**, 891–895.
2. Yusa, K., Horie, K., Kondoh, G., Kouno, M., Maeda, Y., Kinoshita, T., and Takeda, J. (2004) Genome-wide phenotype analysis in ES cells by regulated disruption of Bloom's syndrome gene. *Nature* **429**, 896–899.
3. Luo, G., Santoro, I. M., McDaniel, L. D., Nishijima, I., Mills, M., Youssoufian, H., Vogel, H., Schultz, R. A., and Bradley, A. (2000) Cancer predisposition caused by elevated mitotic recombination in Bloom mice. *Nat Genet* **26**, 424–429.
4. Wang, W., and Bradley, A. (2007) A recessive genetic screen for host factors required for retroviral infection in a library of insertionally mutated Blm-deficient embryonic stem cells. *Genome Biol* **8**, R48.
5. Trombly, M. I., Su, H., and Wang, X. (2009) A genetic screen for components of the mammalian RNA interference pathway in Bloom-deficient mouse embryonic stem cells. *Nucleic Acids Res* **37**, e34.
6. Devon, R. S., Porteous, D. J., and Brookes, A. J. (1995) Splinkerettes-improved vectorettes for greater efficiency in PCR walking. *Nucleic Acids Res* **23**, 1644–1645.

Chapter 5

Construction and Application of Random dsRNA Interference Library for Functional Genetic Screens in Embryonic Stem Cells

Xiaoxing Cheng and Rui Jian

Abstract

RNA interference (RNAi) libraries have been proven to be a powerful tool for large-scale functional genetic screens. To facilitate high-throughput functional genetic screens in embryonic stem cells, a system for construction of random dsRNA expressing RNAi libraries was developed. Previous studies have demonstrated that sequence-specific gene silencing could be induced by long dsRNA in mouse embryos, mouse oocytes, embryonic stem cells, and some other mammalian cells. Our study demonstrated that the dsRNA interference library can be used for functional genetic screens of genes involved in self-renewal of embryonic stem cells (ES cells). The random RNAi library is easy to construct and provides a useful tool for investigation of molecular mechanisms of cellular development and differentiation.

Key words: Functional genetic screen, embryonic stem cells, RNA interference library.

1. Introduction

The first RNA interference (RNAi) library was reported in 2003. It consisted of synthesized siRNA duplexes that targeted 510 individual genes, most of them encoding kinases (1). A systematic screening of modulators of TRAIL-induced apoptosis with this library identified both known and unknown genes. Subsequently, a variety of RNAi libraries consisting of synthesized siRNAs or siRNA vectors that target genes with known sequences or non-coding RNAs have been reported (2–10). Screenings with RNAi libraries have identified new molecules involving many different functions.

B. Zhang, E.J. Stellwag (eds.), *RNAi and MicroRNA-Mediated Gene Regulation in Stem Cells*, Methods in Molecular Biology 650, DOI 10.1007/978-1-60761-769-3_5, © Springer Science+Business Media, LLC 2010

In mouse oocytes and embryos, long dsRNA could induce gene-specific RNAi and the phenotypes exhibited were highly similar to those of null mutants (11, 12). This dsRNA-induced gene-specific knockdown was also observed in ES cells (13–15) and other mammalian cells (13, 16–20). Transgenic expression of long dsRNA targeting *Ski* gene in mouse embryos by using a vector with a modified RNA polymerase II promoter showed phenotypes similar to *Ski*-knockout embryos, suggesting that dsRNA expression did not interfere with normal proliferation and differentiation of early embryo cells, since the normal growth of the embryo was not affected (12). Based on these findings, a novel approach for establishment of cDNA-based random RNAi libraries was developed (**Fig. 5.1**) (5, 21).

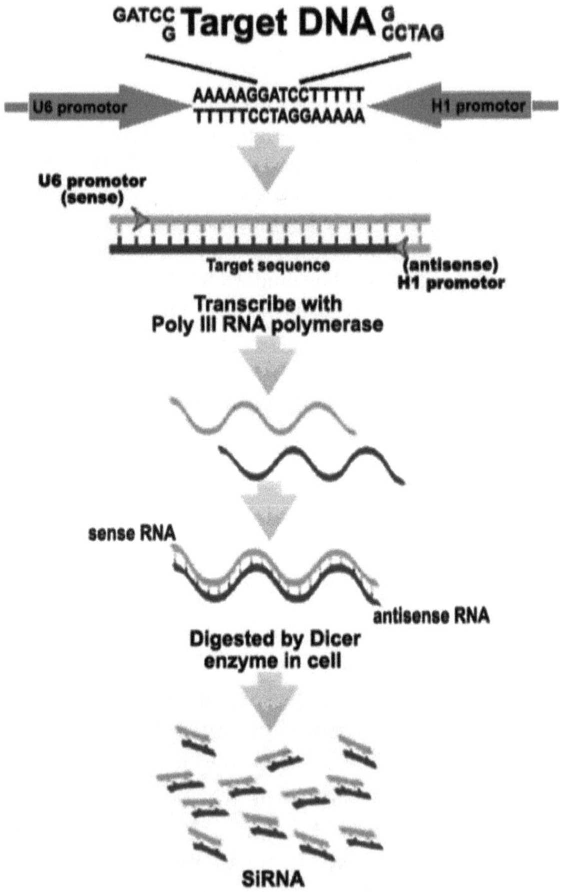

Fig. 5.1. Illustration of the dsRNA interference library. ES cell cDNAs are cloned into the *Bam*HI restriction site between the H1 and U6 promoters. dsRNAs are formed by transcripts from H1 and U6 promoters and are digested by endogenous Dicer to produce siRNA.

2. Materials

2.1. Cell Culture

1. DMEM culture medium: Dulbecco's modified Eagle's medium (DMEM) (Invitrogen, Carlsbad, CA) supplemented with 15% fetal bovine serum (FBS, HyClone, Ogden, UT), 2 mM L-glutamine, 1 mM sodium pyruvate, 0.1 mM nonessential amino acids, and 0.1 mM β-mercaptoethanol (all from Invitrogen).
2. Murine leukemia inhibitor factor (LIF) (Chemicon, Temecula, CA).
3. G418 (Calbiochem, San Diego, CA).
4. All-trans retinoic acid (Sigma, St. Louis, MO).
5. 0.25% trypsin solution: Dissolve 0.25 g trypsin (Sigma) in 100 mL PBS, filter through 0.2-μm membrane, and store at −20°C.
6. 0.1% gelatin: Add 0.1 g gelatin (Sigma) to 100 mL ddH_2O, autoclave at 121°C for 20 min, and store at 4°C.

2.2. Isolation of mRNA from ES cells

1. TRIzol (Invitrogen).
2. PolyATtract mRNA Isolation Systems (Promega, Madison, WI).
3. 20 × SSC stock solution: Combine 175 g of NaCl, 88.2 g trisodium citrate dehydrate, and 800 RNAase-free water. Mix well and adjust pH to 7 with 1 M HCl. Bring volume to 1 L with RNAase-free water. Filter with 0.2-μm filter unit.
4. 0.5 × SSC: To prepare 1.2 mL of 0.5×SSC, add 30 μL of 20×SSC to 1.170 mL of RNAase-free water.
5. 0.1 × SSC: To prepare 1.4 mL of 0.1 × SSC, add 7 μL of 20×SSC to 1.393 mL of RNAase-free water.
6. Sterile RNAase-free water.
7. Magnetic stand (Promega).

2.3. Synthesis of Double-Stranded cDNA

1. Universal RiboClone cDNA synthesis system (Promega).
2. PCR purification kit (Qiagen, Valencia, CA).

2.4. Construction of dsRNA Interference Library

1. *Sau*3A and *Bam*HI (New England Biolabs, Ipswich, MA.)
2. pDoub neo vector (*see* (5)) (**Fig. 5.1**).
3. Phenol/chloroform/isoamyl alcohol (25:24:1).
4. Chloroform/isoamyl alcohol (24:1).
5. 3 M NaAc (pH 5.2).

6. 70% Ethanol.
7. 0.1 cm electroporation cuvette (Bio-Rad, Hercules, CA).
8. LB broth: Tryptone (Difco) 10 g, Yeast extract (Difco) 5 g, NaCl 5 g. Dissolve in distilled water and adjust pH to 7.2 with NaOH, autoclave.
9. LB plates with ampicillin: Containing 100 μg/mL ampicillin (Sigma).

2.5. Transfection of dsRNA Interference Library into ES Cells

1. Plasmid extraction kit (Qiagen)
2. Lipofectamine ™2000 (Invitrogen)
3. DMEM culture medium
4. 0.25% trypsin solution
5. G418

2.6. Identification of Genes Involving in Self-Renewal of ES Cells

1. DNA extraction kit (Qiagen)
2. T7 primer: 5′-GCGCGTAATACGACTCACTATAG-3′
3. KSP primer: 5′-GCTCGAGGTCGACGGTATCGATAAG-3′
4. Taq polymerase, dNTPs (New England Biolabs)
5. PCR purification kit (Qiagen)

3. Methods

The construction of random dsRNA interference library is as follows. First, the RNAi vector pDoub-neo (**Fig. 5.1**) is linearized with *Bam*HI and treated with alkaline phosphatase to prevent self-ligation. Second, cDNAs are digested by *Sau*3AI and DNA fragments ranging from 50 to 500 bps are purified from the gel. Third, the *Bam*HI linearized pDoub-neo vector is mixed with *Sau*3AI-digested cDNAs prepared from stem cells and a ligation reaction is performed. Fourth, the recombinants are transformed into *Escherichia coli*. Fifth, plasmids containing the random RNAi library are extracted from the bacteria.

The plasmids containing the random dsRNA interference library are transfected into ES cells. After identification of desired phenotypes, the EC cell colonies are picked and expanded. The cloned genes are amplified with T7 (5′-GCGCGTAATACGACTCACTATAG-3′)/KSP (5′-GCTCGAGGTCGACGGTATCGATAAG-3′) primers located on the vectors, and genes are identified by sequence analysis.

3.1. Isolation of mRNA from ES Cells

1. Total RNA is extracted from 1 × 10^8 CCE embryonic stem cells with TRIzol reagent, following the protocol provided by the manufacturer (Invitrogen).
2. Add 1 mg of total RNA in 500 μL RNase-free water to a 1.5-mL Eppendorf tube.
3. Heat the tube in a 65°C water bath for 10 min.
4. Add 3 μL biotinylated-oligo(dT) Probe and 13 μL 20×SSC to the tube and mix gently.
5. Let the solution cool slowly to room temperature.
6. Prepare 1.2 mL of 0.5 × SSC: Add 30 μL of 20 × SSC to 1.170 mL of RNAase-free water.
7. Prepare 1.4 mL of 0.1 × SSC: Add 7 μL of 20 × SSC to 1.393 mL of RNAase-free water.
8. Resuspend one tube of streptavidin-paramagnetic particles (SA-PMPs, PolyATtract mRNA Isolation Systems, Promega) by gently flicking the bottom of the tube.
9. Place the tube in the magnetic stand to capture the particles.
10. Carefully remove the supernatant.
11. Wash the SA-PMPs 3 times with 300 μL of 0.5 × SSC.
12. Resuspend the washed SA-PMPs in 100 μL of 0.5 × SSC.
13. Add the entire contents of the annealing reaction from Steps 2–5 to the tube containing freshly washed SA-PMPs.
14. Incubate at room temperature for 10 min.
15. Capture the SA-PMPs by using the magnetic stand and carefully remove the supernatant.
16. Wash the SA-PMPs with 300 μL of 0.1 × SSC.
17. Resuspend the final SA-PMP pellet in 100 μL sterile RNAase-free water.
18. Capture the SA-PMPs by using the magnetic stand.
19. Transfer the eluted mRNA to a sterile RNAase-free tube and store at −70°C freezer until needed.

3.2. Synthesis of Double-Stranded cDNA

1. Add 2 μL of Oligo(dT)$_{18}$ primer (0.5 μg/μL) to 2 μg mRNA in a 1.5-mL sterile tube and bring the volume to 15 μL with RNAase-free water (Universal RiboClone cDNA Synthesis System, Promega).
2. Heat the tube in a 70°C water bath for 10 min and cool on ice immediately.
3. Add 5 μL First Strand 5 × Buffer, 40 U Rnasin ribonuclease inhibitor to the tube.
4. Heat the tube in a 42°C water bath for 3–5 min.

5. Add 2.5 μL sodium pyrophosphate (40 mM) and 30U AMV reverse transcriptase to the tube; bring the volume to 25 μL with RNAase-free water.
6. Incubate the tube in a 42°C water bath for 1 h; store on ice.
7. Add 50 μL second-strand 2.5 × buffer, 6.25 μL acetylated BSA (1 mg/mL), 28.75 U DNA polymerase I, and 1 U RNase H to the tube containing first-strand of cDNA; bring the total volume to 125 μL with RNAase-free water.
8. Mix gently and incubate at 14°C overnight.
9. Heat the tube in a 70°C water bath for 10 min spin briefly, and store on ice.
10. Add 1 μL T4 DNA Polymerase to the tube.
11. Store the tube at 37°C for 10 min.
12. Add 10 μL 200 mM EDTA to stop the reaction and keep on ice.
13. The double-stranded cDNA was purified with DNA purification kit.

3.3. Construction of dsRNA Interference Library

1. Add 5 μL *Sau*3A and appropriate reaction buffer to the tube containing the double-stranded cDNA; bring the volume to 100 μL with sterile ddH_2O.
2. Incubate at 37°C for 2 h.
3. Linearize pDoub-neo vector with *Bam*HI.
4. After purification, mix an equal molar ratio of *Sau*3A-digested cDNA to *Bam*HI-linearized pDoub-neo vector in a 1.5-mL Eppendorf tube; bring the volume to 200 μL with sterile ddH_2O.
5. Add 5 μL T4 DNA ligase (New England Biolabs) to the tube and incubate the reaction at 16°C for 24 h.
6. Add an equal volume of phenol/chloroform/isoamyl alcohol (25:24:1) to the ligation reaction and mix thoroughly.
7. Centrifuge at 12,000 rpm for 5 min and transfer the upper layer to a new tube.
8. Add an equal volume chloroform/isoamyl alcohol (24:1) and mix thoroughly.
9. Centrifuge at 12,000 rpm for 5 min and transfer the upper layer to a new tube.
10. Add 1/10 volume of 3 M NaAc (pH 5.2) and 2.5 volume of ethanol, mix thoroughly, and cool at −20°C for 30 min.
11. Centrifuge at 12,000 rpm for 5 min and discard supernatant.
12. Wash the pellet with 70% ethanol twice.
13. Dry the pellet in the air for 10 min.

14. Dissolve the DNA with sterile ddH$_2$O.
15. Add 1 μL DNA to 50 μL competent *E. coli* DH10B cells in 0.1-cm electroporation cuvette and mix gently.
16. Electroporate using the following settings: 1.8 kV, 2.5 μF, 200Ω.
17. Transfer the cells to a sterile 1.5-mL tube containing 0.9-mL LB, incubated at 37°C for 1 h in a shaker.
18. Spread the bacterial culture on LB Amp plates, 100 μL /plate, incubate at 37°C overnight.
19. Count the number of colonies and calculate the CFU/mL and the titer of the library.
20. Collect the bacteria, suspend in appropriate amount of LB medium, add 0.07% (final concentration) DMSO, aliquot (1 mL/tube) and store at −70°C freezer until needed.

3.4. Transfection of dsRNA Interference Library into ES Cells

1. Thaw one tube (1 mL) of bacteria containing dsRNA interference library.
2. Add 1 mL of bacteria into 100 mL LB broth and incubate at 37°C for 4–6 h with shaking.
3. Extract plasmids from the bacteria using plasmid extraction kit (Qiagen).
4. Transfect plasmids of dsRNA interference library into cultured ES cells by Lipofectamine ™2000 (Invitrogen), following protocol provided by the manufacturer.
5. Culture the ES cells with DMEM culture medium at 37°C in a CO$_2$ incubator overnight.
6. On day 2 after transfection, digest ES cells with 025% trypsin solution, suspend the cells in 10 mL of DMEM culture medium containing 400 μg/mL G418, and add to 10-cm tissue culture dish.
7. Incubate at 37°C in a CO$_2$ incubator for at least 10 days at conditions that induce self-renewal or differentiation.
8. Pick colonies with desired alteration of phenotypes and culture the cells.

3.5. Identification of Genes Involving in Self-Renewal of ES Cells

1. Culture candidate ES cells in DMEM culture medium.
2. Extract DNA from the ES cells using a DNA extraction kit (Qiagen).
3. Amplify cloned genes with T7 (5′-GCGCGTAATACG ACTCACTATAG-3′)/KSP (5′-GCTCGAGGTCGACGGT ATCGATAAG-3′) primers located on the vectors.
4. The PCR reaction is performed by denaturation at 94°C for 1 min, annealing at 60°C for 1 min, and extension at 72°C for 1 min, for a total of 35 cycles.

5. Purify the PCR product with PCR purification kit (Qiagen).
6. PCR product is sequenced and homologues are identified by BLAST search.

3.6. Colony-Forming Assay

1. Digest ES cells with 0.25% trypsin solution to produce a single-cell suspension.
2. Seed approximately 3 × 10^3 cells into gelatin-coated 60-mm tissue culture dish.
3. Incubate the dishes in ES cell culture medium in the presence or absence of 10 ng/mL murine leukemia inhibitor factor (LIF) for 4–6 days at 37°C in a CO_2 incubator.
4. Determine undifferentiated state of ES cells by staining for presence of Oct-4 and alkaline phosphatase.

4. Notes

1. The amount and quality of cDNA are very important for construction of the dsRNA interference library. At least 10 ng of cDNA is needed to construct a library with 1 × 10^5 clones.
2. In the ligation reaction, the ratio of *Sau*3AI-digested cDNA to *Bam*HI-linearized pDoub-neo vector can influence the quality of the library. We recommend using different ratios of cDNA to vector to determine which ratio yields the maximum number of clones.
3. Electroporation generally yields more clones with the same amount of ligated cDNA for transformation of *E. coli*. The ligated DNA should be precipitated and resuspended in milli-Q water without salts.
4. Plasmids derived from the dsRNA interference library can be transfected into ES cells by electroporation; however, many cells die due to electroporation. Transfection with Lipofectamine™ 2000 (Invitrogen) generated the best results and so was adopted for use in this study.
5. The success of the dsRNA interference library screens depends on methods used to identify desired phenotypes. An easy, quick, high-throughput assay should be used to identify ES cell clones that show morphological, biochemical, or functional changes.
6. When positive ES cell clones are picked and expanded, the same phenotype assay should be performed again to confirm the change.

7. When a gene that causes a phenotypic change in ES cells is identified by BLAST search, a new siRNA expressing vector should be constructed based on the sequence of the gene. A phenotype assay should be performed to confirm that the change is indeed caused by knockdown of the gene.
8. A colony-forming assay is used to determine the influence of the gene on self-renewal or differentiation of the ES cells.

Acknowledgments

The work is supported by grants 30370603 and 30470094 from the National Natural Science Foundation of China.

References

1. Aza-Blanc, P., Cooper, C. L., Wagner, K., Batalov, S., Deveraux, Q. L., and Cooke, M. P. (2003) Identification of modulators of TRAIL-induced apoptosis via RNAi-based phenotypic screening. *Mol Cell* **12**, 627–637.
2. Cheng, X., Jian, R., Deng, S., and Jiang, J. (2005) RNA interference library and its application in functional genomics. *Prog Biochem Biophys* **32**, 195–198.
3. Moffat, J., Grueneberg, D. A., Yang, X., Kim, S. Y., Kloepfer, A. M., Hinkle, G., Piqani, B., Eisenhaure, T. M., Luo, B., Grenier, J. K. et al. (2006) A lentiviral RNAi library for human and mouse genes applied to an arrayed viral high-content screen. *Cell* **124**, 1283–1298.
4. Silva, J. M., Li, M. Z., Chang, K., Ge, W., Golding, M. C., Rickles, R. J., Siolas, D., Hu, G., Paddison, P. J., Schlabach, M.R. et al. (2005) Second-generation shRNA libraries covering the mouse and human genomes. *Nat Genet* **37**, 1281–1288.
5. Jian, R., Cheng, X., Jiang, J., Deng, S., Hu, F., and Zhang, J. (2007) A cDNA-based random RNAi library for functional genetic screens in embryonic stem cells. *Stem Cells* **25**, 1904–1912.
6. Kittler, R., Putz, G., Pelletier, L., Poser, I., Heninger, A.K., Drechsel, D., Fischer, S., Konstantinova, I., Habermann, B., Grabner, H. et al. (2004) An endoribonuclease-prepared siRNA screen in human cells identifies genes essential for cell division. *Nature* **432**, 1036–1040.
7. Luo, B., Heard, A. D., and Lodish, H. F. (2004) Small interfering RNA production by enzymatic engineering of DNA (SPEED). *Proc Natl Acad Sci USA* **101**, 5494–5499.
8. Sen, G., Wehrman, T. S., Myers, J. W., and Blau, H. M. (2004) Restriction enzyme-generated siRNA (REGS) vectors and libraries. *Nat Genet* **36**, 183–189.
9. Shirane, D., Sugao, K., Namiki, S., Tanabe, M., Iino, M., and Hirose, K. (2004) Enzymatic production of RNAi libraries from cDNAs. *Nat Genet* **36**, 190–196.
10. Chen, M., Zhang, L., Zhang, H. Y., Xiong, X., Wang, B., Du, Q., Lu, B., Wahlestedt, C., and Liang, Z. (2005) A universal plasmid library encoding all permutations of small interfering RNA. *Proc Natl Acad Sci USA* **102**, 2356–2361.
11. Wianny, F., and Zernicka-Goetz, M. (2000) Specific interference with gene function by double-stranded RNA in early mouse development. *Nat Cell Biol* **2**, 70–75.
12. Shinagawa, T., and Ishii, S. (2003) Generation of *Ski*-knockdown mice by expressing a long double-strand RNA from an RNA polymerase II promoter. *Genes Dev* **17**, 1340–1345.
13. Billy, E., Brondani, V., Zhang, H., Muller, U., and Filipowicz, W. (2001) Specific interference with gene expression induced by long, double-stranded RNA in mouse embryonal teratocarcinoma cell lines. *Proc Natl Acad Sci USA* **98**, 14428–14433.

14. Yang, S., Tutton, S., Pierce, E., and Yoon, K. (2001) Specific double-stranded RNA interference in undifferentiated mouse embryonic stem cells. *Mol Cell Biol* **21**, 7807–7816.
15. Paddison, P. J., Caudy, A. A., and Hannon, G. J. (2002) Stable suppression of gene expression by RNAi in mammalian cells. *Proc Natl Acad Sci USA* **99**, 1443–1448.
16. Park, W. S., Miyano-Kurosaki, N., Hayafune, M., Nakajima, E., Matsuzaki, T., Shimada, F., and Takaku, H. (2002) Prevention of HIV-1 infection in human peripheral blood mononuclear cells by specific RNA interference. *Nucleic Acids Res* **30**, 4830–4835.
17. Gan, L., Anton, K. E., Masterson, B. A., Vincent, V. A., Ye, S., and Gonzalez-Zulueta, M. (2002) Specific interference with gene expression and gene function mediated by long dsRNA in neural cells. *J Neurosci Methods* **121**, 151–157.
18. Yi, C. E., Bekker, J. M., Miller, G., Hill, K. L., and Crosbie, R. H. (2003) Specific and potent RNA interference in terminally differentiated myotubes. *J Biol Chem* **278**, 934–939.
19. Konstantinova, P., de Vries, W., Haasnoot, J., ter Brake, O., de Haan, P., and Berkhout, B. (2006) Inhibition of human immunodeficiency virus type 1 by RNA interference using long-hairpin RNA. *Gene Ther* **13**, 1403–1413.
20. Strat, A., Gao, L., Utsuki, T., Cheng, B., Nuthalapaty, S., Mathis, J. M., Odaka, Y., and Giordano, T. (2006) Specific and nontoxic silencing in mammalian cells with expressed long dsRNAs. *Nucleic Acids Res* **34**, 3803–3810.
21. Jian, R., Peng, T., Deng, S., Jiang, J., Hu, F., An, J., and Cheng, X. (2006) A simple strategy for generation of gene knockdown constructs with convergent H1 and U6 promoters. *Eur J Cell Biol* **85**, 433–440.

Chapter 6

Establishing Efficient siRNA Knockdown in Stem Cells Using Fluorescent Oligonucleotides

Stephen W. Chen and Steve K.W. Oh

Abstract

The following protocols provide a rapid approach for establishing good working conditions for transfecting siRNAs for specific gene knockdown. By first using microscopy to evaluate efficient transfection of an inexpensive, fluorescent oligonucleotide, the researcher can later proceed with more expensive Western blot or quantitative real-time PCR (qRT-PCR) methods. Thus, the main culprit of ineffective knockdown, poor transfection, can be eliminated before engaging in tedious and time-consuming approaches for troubleshooting siRNA knockdown experiments.

Key words: Embryonic stem cells, RNAi, siRNA, transfection, qRT-PCR, gene knockdown, Oct-3/4.

1. Introduction

RNA inhibition (RNAi) is a useful method for studying the function of specific genes in mammalian cells. These cells possess intrinsic mechanisms to target and destroy specific RNA sequences after double-stranded RNA (dsRNA) enters the cell. Even though RNAi is a relatively straightforward method to knockdown specific genes, establishing a reliable and robust method in embryonic stem cells (ESCs) presents several transfection-related obstacles (1, 2).

Some characteristics of ESCs make it difficult to establish effective transfection conditions (3, 4). First, ESCs grow as clustered colonies. This limits the available surface area for siRNA–transfection agent complexes, such as liposome-based reagents.

B. Zhang, E.J. Stellwag (eds.), *RNAi and MicroRNA-Mediated Gene Regulation in Stem Cells*, Methods in Molecular Biology 650, DOI 10.1007/978-1-60761-769-3_6, © Springer Science+Business Media, LLC 2010

Second, ESCs grow rapidly. Rapid growth of cells as clumps exacerbates problems associated with access of transfection agent complexes because of limited surface area. Reducing the cell density limits clumping but can result in cellular differentiation and loss of stem cell activity (5). Seeding the appropriate number of cells for undifferentiated growth must be balanced with sufficient incubation time for transfection complexes to contact the cell surface. Finally, in the presence of transfection agents, ESCs may differentiate and die.

Concentrations of transfection reagents should be minimized to avoid these unwanted effects. Thus, a successful approach to using siRNAs for ESC gene knockdown studies requires (1) seeding enough cells for undifferentiated growth, while maintaining cell surface area; (2) sufficient incubation time; and (3) effective, but minimal, concentration of transfection reagent.

2. Materials

2.1. ESC Cell Culture

1. Mouse ESC lines, E14 and AB2.2.
2. ESC media: High-glucose ES cell media (Gibco), 15% fetal bovine serum (FBS, Hyclone), 1× glutamine/penicillin/streptomycin (Gibco), 143 μM β-mercaptoethanol (Sigma), and 103 U/mL leukemia inhibitory factor (LIF-ESGRO, Chemicon). Store at −20°C and use within 1–2 weeks of preparation.
3. Tissue culture flasks, 12- and 96-well plates (Nunc).
4. Tissue culture gelatin: 0.1% (w/v) Gelatin (Sigma) in water, store at room temperature.

2.2. Transfection and siRNA

1. Commercial transfection reagents, such as Oligofectamine (Invitrogen), Lipofectamine 2000 (Invitrogen), TransFectin (Bio-Rad), Fugene 6 (Roche), Transpass (New England Biolabs), and Exgen (Fermentas).
2. Opti-Mem I (Gibco).
3. 21-mer FAM-labeled oligo dT (light sensitive).
4. siRNAs specific for Oct-3/4:
   "siOct-A" – CCC GGA AGA GAA AGC GAA CUA GCA U
   "siOct-B" – GGU AGA CAA GAG AAC CUG GAG CUU U (Invitrogen Stealth siRNA).
5. FAM-labeled negative control siRNA (Ambion) (light sensitive).

2.3. Reverse Transcription and qRT-PCR

1. 1× PBS.
2. Trizol.
3. Chloroform.
4. MMLV-RT.
5. Oct-3/4 Taqman probe (Applied Biosystems, catalog no. 4331182) (Light sensitive).
6. Gapdh Taqman probe (Applied Biosystems, catalog no. 4308313) (Light sensitive).
7. Taqman PCR Master Mix (Applied Biosystems, catalog no. 4304437) (Light sensitive).

3. Methods

The basic approach is to first use fluorescent microscopy to rapidly screen for effective entry of an inexpensive fluorescent oligonucleotide (FAMdT) into ESC colonies. Using different commercial transfection reagents and varying transfection times after seeding, more effective transfection conditions can be established and used as a starting point for further optimization. On the basis of these observations, one reagent is selected to transfect siRNAs designed to knockdown a specific gene. In this case, Oct-3/4 will serve as an illustrative example.

Following siRNA transfection, RNA is isolated, reverse transcribed, and qRT-PCR is used to measure the levels of gene expression relative to an endogenous control. In this case, an Oct-3/4 Taqman probe (experimental sample) is measured against a gapdh Taqman probe (endogenous control). Once an effective mRNA knockdown method is established, further experiments can be done to isolate both mRNA and protein for qRT-PCR and Western blot analysis.

3.1. Culturing ESCs

1. Maintain ESCs in T25 culture flasks coated with gelatin.
2. To gelatinize a T25 flask, place 6 mL of 0.1% (w/v) sterile gelatin into the flask.
3. Leave the flask overnight at room temperature, and fully aspirate gelatin shortly before passaging cells (*see* **Note 1**).
4. A fully confluent T25 flask contains approximately 20 × 10^6 ESCs and should be passaged following 10-fold dilution.
5. Aspirate cell culture medium, rinse with 3 mL of PBS, then add 1 mL of Trypsin, and incubate at 37°C for 5–7 min.

6. Check flask during incubation under a microscope to ensure cells have properly separated into single cells.
7. While waiting for incubation, add 5 mL of fresh cell culture medium to previously gelatinized T25 flasks.
8. Remove the incubated flask and quench the Trypsin with 9 mL of cell culture medium for a final total volume of 10 mL.
9. Gently flush the cell suspension 1–3× against the flask surface until the suspension is cloudy and homogenous (*see* **Note 2**).
10. Add 1mL of the quenched cell suspension (2×10^6 ESCs) to the 5 mL of culture medium previously aliquoted into the gelatinized T25 flasks for a final culture volume of 6 mL.
11. Passage 1:10 every 3 days and replace with fresh, warm medium on a daily basis.

3.2. Seeding ESCs for Transfection in Either 96- or 12-Wells

1. To seed the ESCs onto a 96-well plate for initial transfection experiments, first gelatinize the tissue culture plate using 100 μL gelatin for coating overnight and aspirate shortly before seeding as described above (*see* **Note 3**).
2. Trypsinize a confluent T25 tissue culture flask as described (*see* **Section 3.1**) and add 15 mL of medium to dilute the 10 mL quenched cell suspension to a final volume of 25 mL (8×10^6 ESCs/mL).
3. Add 100 μL of this diluted cell suspension into a single well of the 96-well plate (8×10^4 ESCs/well).
4. To seed the ESCs onto a 12-well plate for qRT-PCR and Western blot experiments, use 300 μL gelatin for coating (*see* **Note 3**).
5. Add 90 mL of medium to dilute the 10 mL quenched cell suspension to a final volume of 100 mL (2×10^5 ESCs/mL); alternative volumes of 1:10 cell suspension dilutions can be prepared as needed for experiments.
6. From this diluted cell suspension, add 1mL into a single well of the 12-well plate (2×10^5 ESCs/well).

3.3. Transfecting FAMdT into ESCs to Establish Transfection Conditions

1. Transfection of FAMdT can be done according to the respective manufacturers' protocols. To be brief, the protocol for Bio-Rad TransFectin serves as a single, fully illustrated example of transfecting cells seeded in a 96-well plate (*see* **Section 3.2**) (6). Some comments for other reagents are provided as well.
2. Most importantly, transfection *should be performed approximately 4 h after seeding and not after overnight incubation*,

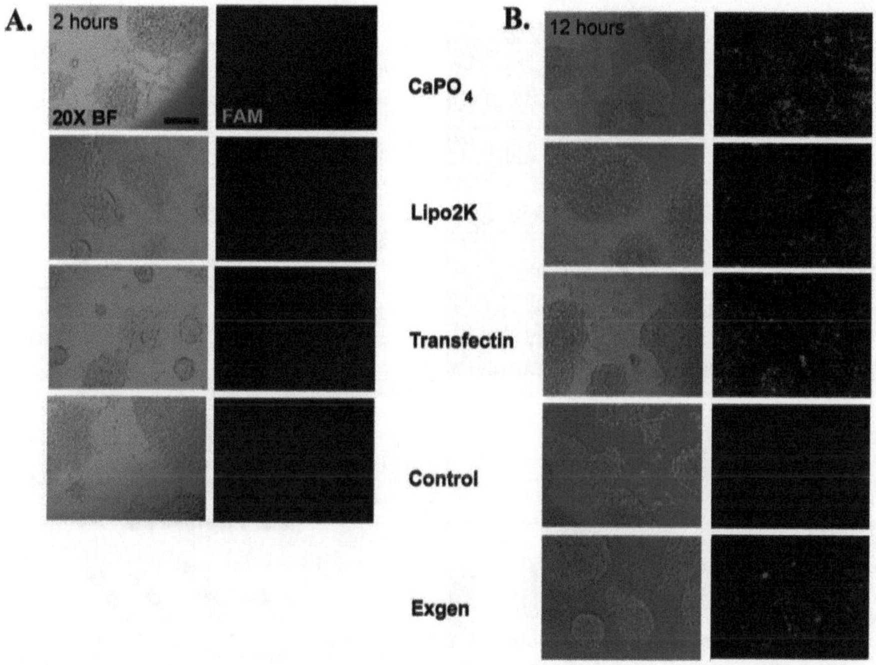

Fig. 6.1. FAM-labeled oligo dT (FAM) transfection in mouse embryonic stem cell colonies seeded overnight and at 40% confluency. Mouse embryonic stem cells, E14 (shown) or AB2.2, were seeded into 96-well plates and visualized at (**A**) 2 and (**B**) 12 h posttransfection for all transfection reagents as indicated (×20 bright field magnification). Scale bar indicates 100 μm. (Reproduced from (7) with permission from Springer.)

contrary to some manufacturers' instructions. Example results are shown in **Figs. 6.1 and 6.2.**

3. Each well is transfected with 3 pmol of FAMdT. For each well, dilute FAMdT in 25 μL of Opti-Mem I serum-free medium. As an example, for the preparation of the exact amount sufficient for 96 wells, 28.8 μL of a 10 μM FAMdT stock solution should be diluted in 2.4 mL of Opti-Mem I serum-free medium (*see* **Note 4**).
4. After diluting the FAMdT, prepare the TransFectin reagent in 25 μL of Opti-Mem I serum-free medium using 0.6 μL TransFectin per well. For a single 96-well plate, mix 57.6 μL of TransFectin into 2.4 mL of Opti-Mem I.
5. Combine the FAMdT and TransFectin solutions. Gently mix by tapping. Incubate for 20 min at room temperature.
6. Add 50 μL of the FAMdT–TransFectin complexes directly to cells in serum-containing medium for a final culture volume of 150 μL.
7. Rock the plate back and forth gently to distribute evenly.
8. Incubate the cells at 37°C in a 5% CO_2 incubator.
9. Use a UV microscope to check the plate at 4-, 8-, and 12-h time points after transfection. Before the first 2-h

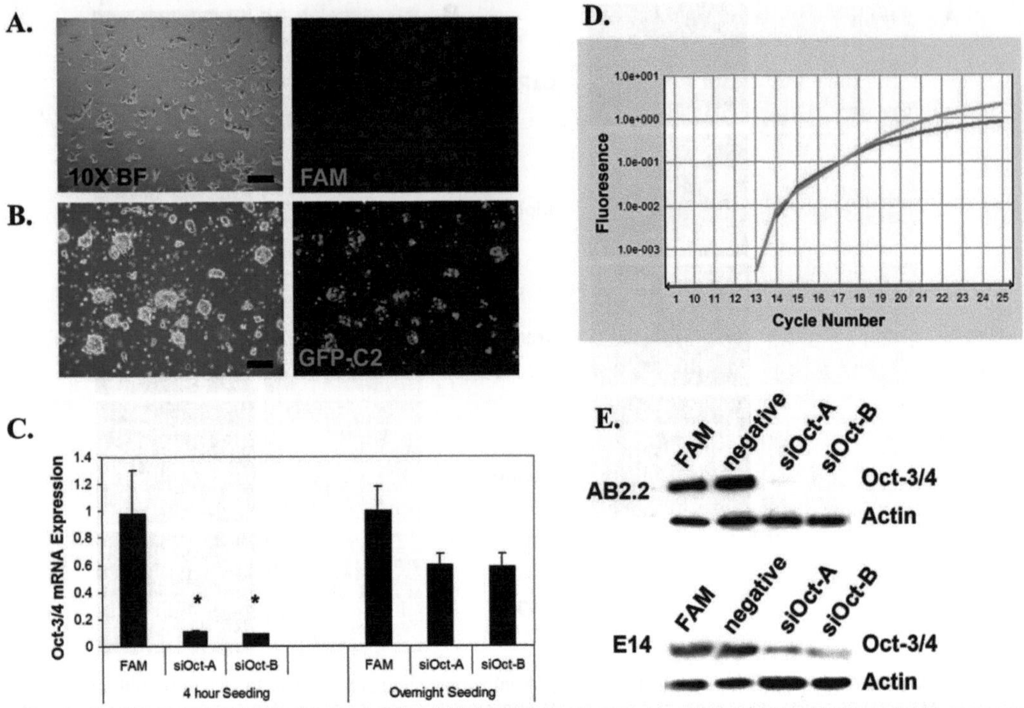

Fig. 6.2. Transfection using small interfering RNAs (siRNAs) and resulting gene knockdown. Rapid transfection 4 h postseeding resulted in high-transfection efficiency using TransFectin and (**A**) FAM-labeled noncognate control siRNA (FAM) or (**B**) GFP-C2 (GFP-C2) expression vector. Scale bar indicates 100 μm (×10 bright field magnificiation). (**C**) Quantitative real-time PCR measurement of Oct-3/4 in E14 mESC using Taqman probes, 24 h posttransfection of FAM, siOct-A, and siOct-B siRNAs as indicated, following 4 h and overnight seeding. Values reported are ΔΔCt normalized to multiplexed endogenous *gapdh* and calibrated against FAM siRNA control with 95% confidence interval as shown ($n = 2$) with * denoting statistical significance ($p > 0.05$). Values for FAM, siOct-A, siOct-B for 2 and 12 h are 0.98, 0.11, 0.09 and 1.00, 0.60, 0.58, respectively. (**D**) Quantitation of *gapdh* amplification (*lower line*) and Oct-3/4 (*higher line*) amplification in a single multiplexed reaction. (**E**) Western blot detection of Oct-3/4 in AB2.2 (*top panel*) and E14 (*bottom panel*) mESC with β-actin (Actin) loading controls shown below. (Oct-3/4, 1:1000 dilution, Santa Cruz and β-actin, 1:2000 dilution, Chemicon). Untransfected mESC (negative), FAM, siOct-A, siOct-B as described, 24 h posttransfection of siRNAs, 4 h postseeding. (Reproduced from (7) with permission from Springer.)

time point, aspirate cell culture medium to remove excess FAMdT before placing under microscope (*see* **Note 5**).

10. In screening for transfection, examine individual wells for fluorescence at both low (×10) and high (×40) magnifications to determine if the fluorescent signal is distributed throughout the clustered cells in the colonies and not restricted to the edges. Switch between bright field and UV light sources to prevent misidentification of bright patches of FAMdT deposited on the tissue culture surface (*see* **Note 6**). An example result is shown in **Fig. 6.1**.

11. Other commercial liposome transfection reagents are added according to the volume specified by the manufacturer for a 96-well plate: Oligofectamine 1 μL,

Lipofectamine 2000, 0.4 μL, Fugene 6, 0.3 μL, Transpass, 2 μL. In all cases, FAMdT and liposomal reagents are complexed in serum-free Opti-Mem I. For nonliposomal reagents: calcium phosphate, 0.36 μL 2 M $CaCl_2$ was added to 3 μL HEPES-buffer solution, Exgen was added to 20 μL 150 mM NaCl solution with a reagent volume of 0.8–1.15 μL (7).

3.4. siRNA Knockdown of Oct-3/4 mRNA

1. Four to 8 h after seeding (*see* **Note** 7) for a 12-well plate as described (*see* **Section 3.2**), ESCs can be transfected for siRNA knockdown.
2. Use multiple designs against a single target gene to eliminate possibility of design-specific RNAi inefficiency or nonspecificity. As an example, two cognate designs for Oct-3/4, siOct-A and siOct-B, are provided (8). In all experiments, a noncognate fluorescent negative control, such as a FAM-labeled siRNA, must be used to visualize successful transfection and measure possible nonspecific siRNA activity.
3. A single 12-well plate is transfected with 30 pmol of siRNA designed to target a specific gene. For each well, prepare 30 pmol siRNA in 100 μL of Opti-Mem I serum-free medium. As an example of preparing the exact amount for 12 wells, 36 μL of a 10 μM siRNA stock solution is diluted in 1.2 mL of Opti-Mem I serum-free medium (*see* **Note 4**).
4. After diluting the siRNA, prepare the TransFectin reagent in 100 μL of Opti-Mem I serum-free medium using 8 μL TransFectin per well. For a single 12-well plate, mix 96 μL of TransFectin into 1.2 mL of Opti-Mem I.
5. Combine the siRNA and TransFectin solutions together by gently tapping the tube to mix. Incubate for 20 min at room temperature. Add 200 μL of the siRNA–TransFectin complexes directly to cells in serum-containing medium for a final culture volume of 1.2 mL.
6. Rock the plate back and forth gently to distribute the siRNA–TransFectin complex evenly.
7. Incubate the cells at 37°C in a 5% CO_2incubator.
8. Before harvesting ESCs, use a UV microscope to screen for transfection of FAM-labeled negative control siRNA. An example result is seen in **Fig. 6.2**.
9. Harvest ESCs 12–24 h after transfection for further analysis, changing cell culture medium at 12 h for longer incubation periods (*see* **Section 3.5**).

3.5. RNA Isolation, Reverse Transcription, and qRT-PCR

1. These instructions largely assume knowledge of general quantitative real-time PCR techniques. Following a 1× PBS

wash, lyse cells with 500 μL Trizol and extract RNA using phenol chloroform extraction.

2. After measuring the RNA using UV spectrophotometry, 4 μg Total RNA is reverse transcribed using MMLV-RT in a 54 μL volume according to the manufacturers' protocol.
3. Use 2 μL of the reverse transcribed cDNA solutions in a single qRT-PCR reaction.
4. Quantitative real-time PCR detection can be done according to manufacturers' protocol with Taqman probes specific for gene knockdown (*see* **Note 8**). Furthermore, an endogenous control for a gene such as *gapdh* can be used as an endogenous control.
5. Reported values are calculated using ΔΔCt method, normalized against endogenous *gapdh* levels, and calibrated against FAM-labeled noncognate control levels (8, 9). An example result is shown in **Fig. 6.2**.

4. Notes

1. For ESC cultures, tissue culture flasks and plates can be more quickly gelatinized in 3 h at 37°C, instead of overnight. However, this is not recommended as routine practice.
2. If colonies are of significantly different size after seeding, make sure to gently flush the cell suspension against tissue culture surfaces using a pipette. This ensures disaggregation into single cells. After seeding, check under the microscope to observe if most cells are properly disaggregated.
3. Cells sometimes aggregate only in the concave middle of the tissue culture well, thereby preventing observation of discrete colonies. When seeding the plate, take care to rock the plate back and forth before placing into the incubator. Never use a circular swirling motion. Another suggestion is to lightly 'drop' the plate in a vertical direction directly onto the incubator shelf. Do not stabilize one edge of the plate and rest the opposite end onto the shelf like a lever. Also, do not shift the position of the plate within the incubator shelf for at least 1 h.
4. For transfection experiments, it is recommended to use glassware, instead of tissue culture polystyrene, to complex nucleic acids and transfection reagents. For complexing light-sensitive nucleotides, wrap in aluminum foil and turn off ambient light sources.
5. If FAMdT signal is weak under UV microscopy, replace ESC medium with serum-free medium, such as Opti-Mem I, to

limit auto-fluorescence. Cells should be in serum-free media only for periods of time needed to observe fluorescence. Longer periods of time may cause cellular differentiations.

6. If further optimization or testing is desired, consider further varying seeding densities and transfection time postseeding. Based on our observations, varying the described oligonucleotide and transfection reagent concentrations did not have a significant effect on transfection efficiency.
7. Extending siRNA transfection incubation periods past 24 h has a negligible effect. RNAi is a very rapid process with the effect mostly occurring within a day. Therefore, extending incubation times is unlikely to improve the results of a knockdown.
8. Instead of Taqman probes, other fluorescent reporter systems such as SYBR Green can be used for qRT-PCR. However, one advantage of Taqman is multichannel fluorescence, allowing multiplexing both target gene and housekeeping gene in a single reaction well.

Acknowledgments

This work is generously supported by funding from the Singapore Agency for Science, Technology and Research (A*STAR).

References

1. Elbashir, S. M., Harborth, J., Weber, K., and Tuschl, T. (2002) Analysis of gene function in somatic mammalian cells using small interfering RNAs. *Methods* **26**, 199–213.
2. Lakshmipathy, U., Pelacho, B., Sudo, K., Linehan, J. L. et al. (2004) Efficient transfection of embryonic and adult stem cells. *Stem Cells* **22**, 531–543.
3. Lorenz, P., Harnack, U., and Morgenstern, R. (2004) Efficient gene transfer into murine embryonic stem cells by nucleofection. *Biotechnol Lett* **26**, 1589–1592.
4. Ma, H., Liu, Q., Diamond, S.L., and Pierce, E.A. (2004) Mouse embryonic stem cells efficiently lipofected with nuclear localization peptide result in a high yield of chimeric mice and retain germline transmission potency. *Methods* **33**, 113–120.
5. Pyle, A.D., Lock, L.F., and Donovan, P.J. (2006) Neurotrophins mediate human embryonic stem cell survival. *Nat Biotechnol* **24**, 344–350.
6. De Zolt, S., Strolz, C., and Altschmied, J. (2005) Highly efficient transfection of mouse ES cells with transfectin lipid reagent. *BioRadiations* **115**, 27–28.
7. Chen, S., Choo, A.B.H., Wang, N.D., Too, H.P., and Oh, S.K.W. (2007) Establishing efficient siRNA knockdown in mouse embryonic stem cells. *Biotechnol Lett* **29**, 261–265.
8. Chen, S., Choo, A.B.H., Wang, N.D., Too, H.P., and Oh, S.K.W. (2007) Knockdown of oct-3/4 and sox-2 attenuates neurogenesis of mouse embryonic stem cells. *Stem Cells Dev* **16**, 413–420.
9. Livak, K. J., and Schmittgen, T. D. (2001) Analysis of relative gene expression data using real-time quantitative PCR and the 2(-Delta Delta C(T)) method>. *Methods* **25**, 402–408.

Chapter 7

Loss-of-Function Studies in Mouse Embryonic Stem Cells Using the pHYPER shRNA Plasmid Vector

Soizik Berlivet, Martin Houlard, and Matthieu Gérard

Abstract

RNA interference is widely used for loss-of-function studies in mammalian cells. As an alternative to the transfection of small RNAs, plasmid vectors have been developed to express short hairpin RNAs (shRNAs). We engineered the pHYPER shRNA vector, which is based on a 2.5-kb mouse genomic fragment encompassing the H1 gene. We have previously shown that this shRNA vector is highly efficient for both transient transfection studies in embryonic stem (ES) cells and generation of stable ES cell lines. Following ES cell transfection, the H1 promoter of pHYPER is recognized by the RNA polymerase III machinery, which directs the transcription of the shRNA. We provide here detailed protocols that explain how to optimize the use of pHYPER in ES cells.

Key words: RNA interference, shRNA, embryonic stem cell, pHYPER, GFP, p150 CAF-1.

1. Introduction

RNA interference has been used in a wide range of organisms to analyze gene function. It was adapted to mammalian cells following the discovery that transfected small double-stranded RNAs can target the degradation of specific RNAs (1). Plasmid vectors that transcribe short hairpin RNAs (shRNAs) have been developed and successfully applied to suppress gene expression in a wide series of cell lines (2–7). We have set up a new shRNA plasmid vector for use in mouse embryonic stem (ES) cells. This vector, pHYPER, allows robust expression of shRNAs from a mouse H1 promoter (8). pHYPER is efficient for the generation of stable ES cell lines expressing an shRNA. In addition, this vector

B. Zhang, E.J. Stellwag (eds.), *RNAi and MicroRNA-Mediated Gene Regulation in Stem Cells*, Methods in Molecular Biology 650, DOI 10.1007/978-1-60761-769-3_7,

can also be used for the knockdown of genes whose products are essential for cell viability. We describe here the protocol that we set up to target the degradation of the p150 subunit of CAF-1 (Chromatin Assembly Factor 1), which is required for ES cell growth and self-renewal (9). We analyzed CAF-1 depletion by Western blot and immunofluorescence.

2. Materials

2.1. Vector Construction

2.1.1. Preparation of the Oligonucleotides Encoding the shRNA

1. T4 polynucleotide kinase (NEB, Ipswich, MA).
2. T4 polynucleotide kinase buffer (NEB, Ipswich, MA).
3. ATP 10 mM (GE Healthcare) (*see* **Note 1**).

2.1.2. Oligonucleotides Cloning

1. Restriction enzymes: EcoRV, KpnI, MluI, and XhoI (Invitrogen); BstBI (NEB, Ipswich, MA).
2. Digestion Buffer: Invitrogen React® Buffer 1, 2, and 3; NEB Buffer 4 (NEB, Ipswich, MA).
3. Agarose (Invitrogen).
4. Nucleospin® Extract II (Macherey-Nagel).
5. T4 DNA Ligase Buffer (5 mM ATP) 10x (Invitrogen).
6. T4 DNA Ligase (Invitrogen).
7. *Escherichia coli* chemically competent bacteria.
8. LB medium (without and with 75 μg/mL ampicillin).
9. LB Agar Petri dishes with 75 μg/mL ampicillin.
10. Solution I (4°C): Glucose 50 mM, Tris/HCl 25 mM pH 8.0, EDTA 10 mM.
11. Solution II (room temperature): NaOH 200 mM, SDS 1%.
12. NaOAc 3 *M*, pH 5.2.
13. TE (Tris 10 mM, PH 7.5, EDTA 1 mM).
14. NucleoBond PC 500® Plasmid kit (Macherey Nagel).
15. Phenol-chloroform (Fluka; Phenol is toxic. Chloroform is an irritant that is harmful and may cause reproductive damage. Hence, great care should be taken when handling these products).
16. Ethanol 100 and 70%.

2.2. Cell Culture and Electroporation

1. Dulbecco's Modified Eagle's Medium (DMEM) (Invitrogen).
2. Fetal calf serum (Invitrogen).
3. 0.25% trypsin solution (Invitrogen).
4. 200 mM Glutamine (Invitrogen).
5. PBS 1x : 137 mM NaCl, 2.7 mM KCl, 4.3 mM Na_2HPO_4, 1.47 mM KH_2PO_4, pH 7.4.
6. 100x nonessential amino acids (Invitrogen).
7. Leukemia Inhibitor factor (LIF, Euromedex).
8. Penicillin/streptomycin (Invitrogen).
9. β-mercaptoethanol (Sigma; very hazardous in case of skin contact, ingestion, eye contact, or inhalation. Caution should be taken to avoid exposure).
10. D10 medium: DMEM with 10% fetal bovine serum.
11. D15 medium: 15% fetal bovine serum, 50 U/mL of penicillin/streptomycin, 0.1 mM nonessential amino acids, 0.1 mM β-mercaptoethanol, 10^7 U/mL LIF (leukemia inhibitor factor), complete with DMEM.
12. 1% gelatin solution in water (Sigma).
13. Pasteur pipettes.
14. Electroporation cuvettes (gap width 0.4 cm, Bio-Rad).
15. Microscope glass cover slips 18×18 (VWR).

2.3. ES Cell Selection

1. D15 with 2 μg/mL puromycin (Sigma).
2. PBS 1x.

2.4. Immunofluorescence

1. Paraformaldehyde 4%.
2. PBS 1x.
3. Permeabilization buffer: PBS 1x, 1% Triton®X-100 (Sigma).
4. Saturation solution: PBS 1x, 0.1% Tween 20, 1% gelatin (Sigma).
5. Primary antibody: p150 CAF-1 antibody (9).
6. Secondary antibody: Alexa Fluor 594 goat anti – rabbit IgG (Invitrogen).
7. Di aminido phenyl Indol (DAPI).
8. Vectashield (Vector Laboratories, Burlingame, CA).

2.5. Western Blot

2.5.1. Protein Preparation

1. PBS 1x.
2. 0.25% trypsin solution.

3. D10 culture medium.
4. Lysis buffer: 2 mM Hepes, pH 7.9, KCl 500 mM, Glycerol 20%, Igepal 0.1% (Sigma), DTT 1 mM, Complete (Roche).
5. Bio-Rad Protein Assay.
6. Laemmli buffer (2x): 4% SDS, 20% glycerol, 10% β-mercaptoethanol, 0.004% bromphenol blue, 0.125 *M* Tris HCl, pH 6.8.
7. β-mercaptoethanol.

2.5.2. SDS-Polyacrylamide Gel Electrophoresis (SDS-PAGE) and Transfer

1. β-mercaptoethanol.
2. BenchMark^{TM} Pre-Stained Protein Ladder (Invitrogen).
3. Acrylamide/bisacrylamide (30:0.4) and (20:0.56).
4. Tris 1 M, pH 8.8 and Tris 1 M, pH 6.8.
5. Sodium dodecyl sulfate (SDS) 20% (Invitrogen).
6. Ammonium persulfate (APS) 10% (w/v, Sigma).
7. TEMED (Sigma).
8. Bromophenol Blue (Sigma).
9. Running buffer: 25 mM Tris base, 192 mM Glycine, 0.1% SDS.
10. Transfer buffer: 4.8 mM Tris base, 38 mM Glycine, 0.04% SDS, 0.2% ethanol.
11. Nitrocellulose membrane (Whatman).
12. TBS 10x: 250 mM Tris-HCl, pH 7.5, 1.37 M NaCl, 27 mM KCl.
13. Blocking solution: TBS1x, Tween 0.1%, skim milk 20%.
14. Primary antibody: p150CAF-1 antibody (10).
15. Secondary antibody: anti-rabbit IgG HRP, dilution 1/2000 (Promega).
16. Washing buffer: TBS1x, Tween 0.3%, skim milk 20%.
17. ECL (Pierce).

3. Methods

pHYPER is a 6,760 bp long plasmid that drives the expression of an shRNA against GFP mRNA (8). This vector harbors the ampicillin-resistance gene for selection of bacteria and a puromycin-resistance cassette for selection of mammalian cells in culture. A schematic representation of the pHYPER vector is given in **Fig. 7.1**. New sequences encoding shRNAs can be intro-

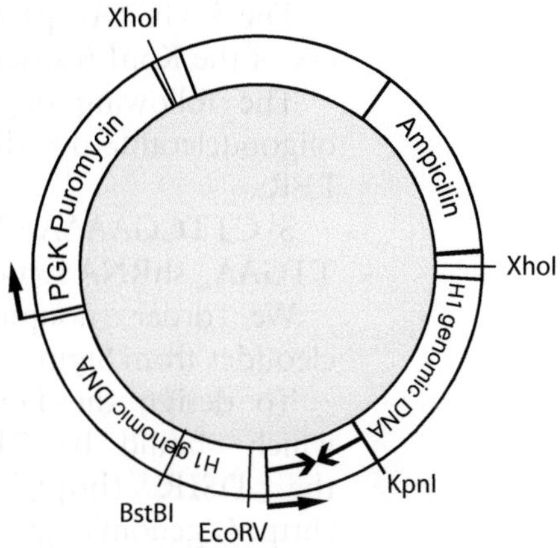

Fig. 7.1. Map of the pHYPER shRNA plasmid. This vector contains a 2.5-kb H1 promoter and a puromycin-selection cassette in the opposite orientation with respect to H1 transcription. PGK: Phospho-glycerate-kinase promoter, Puro: puromycin-resistance gene, AMP: ampicillin-resistance gene.

duced between the EcoRV and KpnI restriction sites. The construction of a new pHYPER vector targeting an endogenous protein is usually performed in a single step (*see* **Note 2, 3, and 4**). We generated the pHYPER@p150CAF-1 that targets the degradation of the mRNA encoding the p150 subunit of Chromatin Assembly Factor 1 (CAF-1) (9).

Depletion of the targeted protein must be analyzed carefully. We recommend, when possible, to use both immunofluorescence and Western blot to confirm the knockdown.

3.1. DNA Construct

3.1.1. Design of the shRNA Coding Oligonucleotide

1. New sequences encoding shRNAs are introduced into pHYPER by ligation of double-strand oligonucleotides with the following design:

   5′ ATCCTGAA_ shRNA sense strand sequence _TTCA AGAGA_ shRNA antisense sequence_TTTTTTCGAA GGTAC **3**′.

   The ATCCTGAA sequence (position +1; +8) is part of the H1 promoter and must be added in every new oligonucleotide construct. When cloned into EcoRV (blunt) digested DNA, it will allow the conservation of the EcoRV restriction site. The TTTTTT sequence is a transcription termination signal for RNA polymerase III.

   The TTCGAA sequence downstream of the termination signal is a BstBI restriction site that we use to screen plasmid minipreps after cloning of the shRNA-encoding oligonucleotides.

The 3′ GGTAC part of the sequence will allow the recovery of the KpnI restriction site.

The following oligonucleotide is hybridized with the oligonucleotide described above before ligation into pHYPER:

5′ CTTCGAAAAAA_ shRNA antisense sequence _TCTC TTGAA_ shRNA sense strand sequence _TTCAGGAT 3′

We order nonpurified, unphosphorylated oligonucleotides from Sigma.

To design the DNA sequence encoding the shRNA, which should be 21 nt long, we recommend either the DSIR (http://cbio.ensmp.fr/dsir/) or SiDirect (http://genomics.jp/sidirect/) software.

To target p150CAF-1, we selected the following sequence:

5′ AAGGAGAAGGCGGAGAAGCAG 3′

2. Phosphorylation of the oligonucleotides is required prior to the ligation. In a reaction volume of 25 μL, pipet 100 pmoles of oligonucleotide, 2.5 μL of 10x T4 polynucleotide kinase buffer, 2.5 μL of 10 mM ATP, 10 units of T4 polynucleotide kinase, and incubate for 30 min at 37°C.
3. When the phosphorylation is completed, mix the two complementary oligonucleotides and transfer the tube into a water bath at 90°C. Switch off the heating system of the water bath and let the temperature of the water slowly cool to room temperature. The two complementary oligonucleotides will anneal during this step.

3.1.2. Ligation of the Oligonucleotide Encoding the shRNA into pHYPER

1. We cloned the double-stranded DNA oligonucleotides encoding the shRNA into the EcoRV and KpnI restriction sites of pHYPER according to the following strategy. 5 μg of vector is first digested using 10 units of KpnI in a 30 μL reaction volume. After 1 h at 37°C, the reaction volume is adjusted to 100 μL in EcoRV digestion buffer, 10 units of EcoRV are added, and digestion is performed for one more hour at 37°C. The digested fragment, which is 6,725 bp long, is purified using the NucleoSpin® Extract II kit.
2. For the ligation, mix 0.1 pmole of the purified, digested pHYPER vector (420 ng) and 0.3 pmol of phosphorylated, hybridized oligonucleotides and bring the volume to 16 μL with water. Add 2 μL of T4 DNA ligase buffer 10x (containing 10 mM ATP) and 2 units of ligase. Incubate at room temperature for 12 h.
3. Thaw 100 μL of chemically competent *E. coli* bacteria on ice. Add 5 μL of the ligation product and mix gently. Incubate the bacteria on ice for 20 min. Heat-shock the bacteria

at 42°C for 30 s. Place them on ice for 1 min. Add 200 μL of LB medium and incubate the bacteria at 37°C for 30 min. Spread the bacteria on Petri dishes containing LB Agar with ampicillin (75 μg/mL) (*see* **Note 5**).

4. The following day, pick up 12 colonies and culture them overnight in 3 mL of LB medium with ampicillin (75 μg/mL).
5. The next day, collect 1.5 mL of each culture into 1.5-mL microfuge tubes and centrifuge the bacteria (30 s at 13,000 rpm). Discard the supernatant and resuspend the pellet in 100 μL of solution I. Add 200 μL of solution II and incubate for 5 min at room temperature. Add 150 μL of NaOAc 3 M, pH 5.2. Mix by inversion and then centrifuge for 10 min in a microcentrifuge at 13,000 rpm. Transfer the supernatant to a new tube and add 1 mL of EtOH 100%. Incubate for 20 min on ice to precipitate DNA and then centrifuge for 10 min at 13,000 rpm at 4°C. Discard the supernatant and wash the pellet with 750 μL of EtOH 70% (centrifuge for 5 min at room temperature at 13,000 rpm in a microcentrifuge). Dry the pellet and then resuspend it in 50 μL of TE buffer.
6. The restriction enzyme B*st*BI is used to check for positive clones. 5 μL of plasmid miniprep (0.1 μg/μL) is used for digestion in a final volume of 20 μL. Add 2 μL of 10x NEB buffer 4 (1x final), 3 units of BstBI, and incubate at 65°C for 1 h. Add 2 μL of loading buffer and check the digestions on a 1% agarose gel. Positive clones are identified by the presence of two DNA fragments of 6.2 and 0.5 kb, while the parental plasmid is linearized by *BstB*I but does not display the 0,5 kb fragment.
7. The positives clones are sequenced with the following primer:
   5′TTCCAGAGCCTGATCTCT 3′.
8. Plasmid amplification: Culture one positive clone in 100 mL of LB media with ampicillin (75 μg/mL). Extract and purify the plasmid with a NucleoBond PC 500® Plasmid kit according to the manufacturer's protocol.

3.2. ES Cell Culture and Electroporation

We used LTM7 derived from (C57BL/6x129) F1 females bred with C3H/HeJ males. The ES cells are cultured on a layer of embryonic fibroblasts in D15 medium.

The cells are ready for electroporation when the culture dishes reach about 80% confluency.

1. In the morning, prepare the cover slips that will be used for immunofluorescence. Wash a series of microscope cover slips with ethanol and place them onto a 10-cm tissue cul-

ture dish. Once completely dried, transfer 4 cover slips to a new 10-cm dish and add 5 mL of 1% gelatin solution. Gelatin coating is required for ES cell growth onto cover slips. Discard the gelatin solution after 1 h. The cell culture dish is ready for use. ES cells will be directly seeded onto it after electroporation (*see* **Note 7**).

2. Change the D15 medium of the cultured cells 4 h before electroporation (*see* **Note 8**).
3. To prepare the cells for electroporation, first discard the D15 medium.
4. Wash each 10-cm ES cell culture dish once with 10 mL of warm (37°C) PBS 1x. Remove the PBS before proceeding to step 5.
5. Cover the cells with 2 mL of 0.25% trypsin solution and incubate for 3 min at 37°C.
6. Add 5 mL of D10 to neutralize the trypsin.
7. Pipet several times to obtain a single cell suspension.
8. Transfer the cells to a 50-mL centrifuge tube.
9. Count the cells.
10. Centrifuge the cells at 1,000 rpm for 5 min.
11. Aspirate the supernatant and resuspend the cells in 50 mL of PBS 1x and centrifuge one more time at 1,000 rpm for 5 min.
12. Aspirate the supernatant and resuspend the cells in PBS 1x to reach a cell density of 2.5×10^7 cells per mL.
13. Transfer 0.8 mL of the cell suspension to an electroporation cuvette.
14. Add 20 μg of pHYPER vector to the cells. Mix the cells and the plasmids with a plastic Pasteur pipette (*see* **Note 9**).
15. Electroporate the cells at 250 V and 500 μF, using a Gene Pluser II Bio-Rad electroporator.
16. Transfer the cells with a plastic Pasteur pipette to a 50-mL centrifuge tube containing 30 mL of D15 (*see* **Note 10**) and aliquot 3 × 10 mL of the resultant cell suspension to cell culture dishes (10 ml per dish).

3.3. ES Cell Selection

1. 24 h after the electroporation, wash the cells with 10 mL of PBS 1x. Aspirate the supernatant and add 10 mL of D15 with puromycin (2 μg/mL).
2. The next day, cultured cells are washed with 10 mL of PBS 1x in order to remove dead cells, supernatant aspirated and replaced with 10 mL of D15 containing puromycin

(2 μg/mL). This process is repeated daily for at least 48 h. This period is necessary for proper selection by puromycin (*see* **Note 11**).

3.4. Immunofluorescence

ES cells electroporated with the pHYPER@CAFp150 were analyzed by immunofluorescence (**Fig. 7.2**).

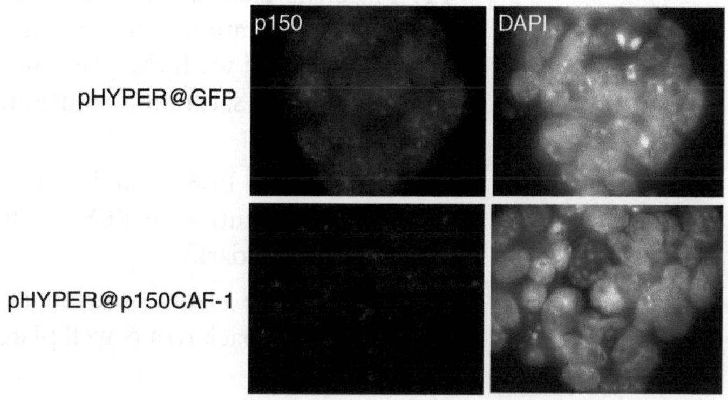

Fig. 7.2. Knockdown of an endogenous gene using pHYPER in a transient transfection protocol. Immunodetection of p150 CAF-1 in ES cells transfected with a control shRNA that target GFP (pHYPER@GFP) or p150CAF-1.

1. After 48 h of puromycin selection, wash carefully the cell culture dish dedicated to immunofluorescence twice with 5 mL of PBS 1x.
2. Transfer each of the four glass slides to a well of a 6-well plate containing 3.5 mL of PBS 1x.
3. Wash the glass slides one more time with PBS 1x. Discard the PBS.
4. Fix the cells with PFA 4% (3.5 mL/well) at room temperature for 20 min (*see* **Note 12**).
5. Wash the glass slides three times with PBS 1x.
6. Permeabilize the cells with 3.5 mL of PBS 1x, Triton® X-100 1% at room temperature for 10 min.
7. Wash the glass slides three times with PBS 1x.
8. Place the glass slides in 3.5 mL of a saturation solution (PBS 1x, 0.1% Tween, 1% gelatin) at 4°C for 20 min.
9. Apply the primary antibody to the glass slides for 1 h at room temperature. To minimize the volume of antibody, a 50-μL drop of a 1/250 antibody dilution is placed onto a square of Para-Film (one for each glass slide). Place each glass slide over a drop of the antibody solution with the cells facing the solution.
10. Return the glass slides to a 6-well plate containing 3.5 mL of saturation solution for 5 min at 4°C. After incubation,

remove the solution and wash the glass slides two more times with fresh saturation solution.

11. Apply a 1/500 dilution of the secondary antibody to the glass slides for 1 h at room temperature, in the dark. The same system as described above is applied to minimize the volume of secondary antibody.
12. Place the glass slides back to a 6-well plate containing 3.5 mL of saturation solution for 5 min at 4°C. Discard the solution and wash the glass slides two more times with 3.5 mL of fresh saturation solution, then once with 3.5 mL of PBS 1x.
13. Remove the PBS 1x and incubate the glass slides in DAPI (1/5000 dilution in PBS 1x) for 2 min at room temperature (in the dark).
14. Remove the DAPI solution by aspiration, then place the glass slides back to a 6-well plate containing 3.5 mL of PBS 1x for wash.
15. The glass slides are then mounted (*see* **Note 13**). Pipet 10 μL of Vectashield onto a microscope glass slide. Apply the cover slip to the Vectashield (cells facing the slide).
16. The glass slides are then sealed with nail polish, which is initially applied to the corners of the glass slides, which are dried and further treated with nail polish until the entire perimeter of the coverslip has been sealed.

3.5. Western Blot Analysis of Protein Depletion

Western blot quantification of protein abundance after knockdown is the most sensitive method, and therefore should be used whenever a good antibody is available (**Fig. 7.3**).

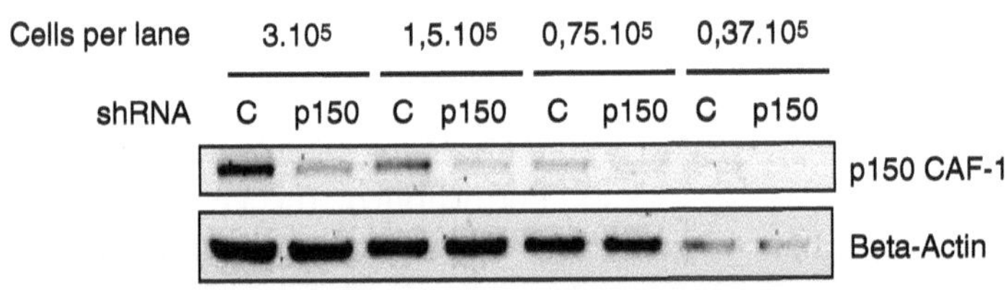

Fig. 7.3. Western blot analysis of ES cells transfected with a pHYPER vector that target p150CAF-1 or a control sequence (c). The number of cells used in each lane is indicated. The amount of p150CAF-1 is significantly reduced in ES cells transfected with pHYPER@p150CAF-1 compared with the control.

3.5.1. Protein Extract

We prepare a protein extract from the puromycin-resistant cells cultivated in the 10 cm cell culture dish according to the following steps.

1. At the end of the puromycin selection, remove the D15 medium and wash the confluent ES cell-containing dish with 10 mL of 1x PBS.

2. Cover the cells with 2 mL of 0.25% trypsin solution and incubate for 3 min at 37°C and then add 5 mL of D10 to neutralize the trypsin.
3. After trypsinization, pipet several times to make single cell suspension. Transfer the cells to a 50-mL centrifuge tube.
6. Count the cells and centrifuge at 1,000 rpm for 5 min.
7. Wash the cells in 10 mL of 1X PBS and centrifuge at 1,000 rpm for 5 min. Repeat this washing step.
8. Lyse the cells by adding 100 μL of lysis buffer per 2.10^6 cells. Homogenize by pipeting up and down.
9. The lysate is then sonicated (power 5 for 10s) with a Microson ultrasonic cell disruptor (Misonix). The lysate is transferred to a 1.5-mL centrifuge tube.
10. Centrifuge for 10 min at 12,000 rpm at 4°C to pellet the cell debris.
11. Transfer the supernatant (containing the proteins) to a new 1.5-mL tube. Store at –80°C if the samples are not analyzed the same day.
12. The proteins extracted are quantified using a Bradford assay.
13. Adjust the concentration of protein in the different samples to 2 μg/μL.
14. For the Western blot, thaw on ice 10 μL of each sample (approximately 20 μg of proteins). Add 12.5 μL of Laemmli Buffer 2x and 2.8 μL of β-mercaptoethanol to a final volume of 25 μL.
15. Incubate for 5 min at 95°C. The samples are ready for loading onto a polyacrylamide gel.

3.5.2. Western Blot

1. The following procedure for western blotting is based on the use of a Hoefer SE260 Mighty Small II gel system for sample PAGE electrophoresis.
2. Prepare a 1-mm thick, 8% gel by mixing 1.6 mL acrylamide/bis solution (30:0.4) with 1.5 mL of 1 *M* Tris-HCl, pH 8.8, 2.826 mL of water, 30 μL of SDS 20%, 40 μL of APS 10%, and 4 μL of TEMED. Pour the gel, leaving space for a stacking gel, and overlay with distilled water. The gel polymerizes in about 30 min (*see* **Note 14**).
3. Pour off the water.
4. Prepare the stacking gel by mixing 388 μL of acrylamide/bis solution (20:0.56) with 250 μL of Tris 1.5 *M* pH 6.8, 1.325 mL of water, 20 μL of SDS 20%, 40 μL of APS 10%, 6 μL of TEMED, and 6 μL of bromophenol

blue. Pour the gel and insert the sample well-former. The stacking gel should polymerize within 30 min.

5. Prepare the running buffer by combining 20 mL of the 10x running buffer with 180 mL of water.
6. Once the stacking gel has polymerized, carefully remove the sample well-former and wash the wells with running buffer.
7. Add the running buffer to the upper and lower chambers of the gel unit. Load the samples, including one well for a prestained molecular weight marker.
8. Complete the assembly of the gel unit and connect to the power supply. Run the gel at 15 mA for 1 h.
9. After electrophoresis, the proteins are transferred electrophoretically to a nitrocellulose membrane. The following instructions are based on the use of an Electro-blotting B33 apparatus (Biometra). Prepare three sheets of 3 MM Chr, Whatman paper larger than the separating gel. Lay them on the lower part of the electro-blotting apparatus. Add transfer buffer to wet the 3 MM sheets. Cut a sheet of nitrocellulose the same size as the separating gel. Lay the membrane on the 3 MM papers (*see* **Note 15**).
10. The gel unit is disconnected from the power supply and disassembled. First the gel is removed from the glass plates and then placed on the nitrocellulose membrane, which is sandwiched between Whatman 3MM paper. After ensuring that the bubbles between the gel and the nitrocellulose and the Whatmann 3MM paper have been removed, the entire sandwich is placed between the two electrode plates of the transfer apparatus and secured to ensure firm contact between the gel and nitrocellulose is maintained throughout the transfer process.
11. Transfer is performed at 120 mA for 45 min.
12. Once the transfer is completed, the upper part of the transfer apparatus and the 3 MM papers are removed.
13. The nitrocellulose membrane is then incubated in 50 mL of blocking buffer for 30 min at room temperature on a rocking platform.
14. The blocking buffer is discarded and the membrane is placed in 10 mL of blocking solution containing a 1/500 dilution of the p150 CAF-1 antibody. The membrane is incubated for 1 h at room temperature.
15. The primary antibody is removed and the membrane washed three times for 5 min with 50 mL of blocking solution.

16. The secondary antibody is freshly prepared for each experiment as a 1/5000 fold dilution (in 10 mL of blocking buffer) of the antibody stock solution purchased from the manufacturer. The membrane is incubated at room temperature for 1 h on a rocking platform.
17. The secondary antibody is discarded and the membrane is washed three times with 1x TBS, 0.3%, Tween, 20% skimmed milk, and then three times with blocking solution.
18. Once the final wash is removed from the blot, the ECL reagent is prepared according to the manufacturer's instructions and then immediately added to the blot. The membrane with ECL is placed in a dark place for 1 min.
19. The membrane is removed from the ECL reagent, blotted with Kim-wipes, and placed onto an acetate sheet protector in an X-ray film cassette with a film (Amersham Hyperfilm ECL, GE Healthcare). Usually, a 2-min exposure time is appropriate (*see* **Note 16**).

4. Notes

1. Unless stated otherwise, all solutions should be prepared in water that has a resistivity of 18.2 MΩ-cm and total organic content of less than five parts per billions. This standard is referred to as 'water' in this text.
2. pHYPER has been validated in ES cells. We recommend to first test its efficiency when using other cell types.
3. We routinely used the pHYPER@GFP vector as a control in our knockdown experiments.
4. A two-step cloning strategy can be used if the one-step cloning is unsuccessful. In this strategy, the first step allows the introduction of the shRNA sense strand. The second step introduces the shRNA antisense strand.

   We recommend adding restriction sites to each of the sense and antisense strands to facilitate the cloning. A TTTTT 'STOP' signal for RNA polymerase III following the antisense strand should also be introduced. The *BamHI* restriction site can be used as a loop between the sense and antisense strands of the shRNA.

   The sense strand has the following structure.

   shRNA sense strand: *EcoRV*shRNA sense strand *BamHI-NotI-KpnI*

The oligonucleotides ordered for the shRNA sense strand have the following design:

5′ATCCTGAA5′ shRNA sense strand-3'GGATCCGCG GCCGCGGTAC 3′

3′TACGACTT3′ shRNA sense strand5′CCTAGGCGC CGGCGC 5′

The antisense strand has the following structure.

shRNA antisense strand: BamHIshRNA antisense strandSTOP-BstBI-KpnI

The oligonucleotides ordered for the shRNA antisense strand have the following form:

5′GATCC5′ shRNA antisense strand3′TTTTTTTC GAAGGTAC 3′

3′CTAGG3′ shRNA antisense strand5′AAAAAAA GCTT C5′

5. The cloning of the oligonucleotide is easy. We usually obtained more than 100 bacterial colonies per Petri dish.
6. We usually obtained more than 10 positive over 12 clones.
7. Once treated with the gelatin solution, the glass slides should be used in the following hours. They would otherwise stick to the dish.
8. All solutions (PBS, D10, D15, and trypsin) are warmed and used at 37°C to avoid heat-shock reactions of the ES cells.
9. For transient transfection, the pHYPER vector does not need to be linearized. For the generation of stable clones, the pHYPER vector should be digested with XhoI and agarose gel-purified in order to remove the plasmid-related DNA sequences. Mix 20 μg of vector, 5 μL of Invitrogen React® Buffer 2, and 20 U of XhoI in a final volume of 50 μL. Incubate for 2 h at 37°C. The digested plasmid is then electrophoresed in a 0.8% agarose gel, and the 4,480-bp long fragment is selected. It is gel extracted and purified using the NucleoSpin® Extract II kit.
10. During electroporation, a proportion of ES cells are lysed and DNA is released, which makes the electroporated cell suspension viscous.
11. For new shRNAs, we recommend testing different time points. Depending on the half-life of the protein, the inhibition may be apparent as soon as 48 h or later. To obtain stable clones expressing the shRNA targeting GFP, the puromycin selection was performed over a period of 5 days. For the CAF-1p150 knockdown study, the puromycin selection was restricted to 48 h, because of massive cells death after 72 h of knockdown. Puromycin

selection should be performed at least 48 h to remove all untransfected cells.

12. We usually prepare 8% paraformaldehyde in 1x PBS that we store at −20°C for up to 1 month. The 4% dilution is prepared just before use.
13. Air bubbles are undesirable in the mounting medium. Slow, careful application of the top layer minimizes their appearance.
14. The percentage of acrylamide of the gel depends on the molecular weight of the proteins. Here we used an 8% acrylamide gel which is adapted to the p150 CAF-1 protein.
15. It is important that the blotter paper (and the membrane) is not larger than the gel. Larger pieces could make the transfer inefficient.
16. We placed a luminescent tape on the acetate sheet to align the film and the membrane. It allows a precise identification of the signals with the lanes of the gel.

Acknowledgments

The authors thank Sylvie Jounier and Hélène Humbertclaude for the production of feeder cells for ES cell culture. This work was supported by the Association pour la Recherche sur le Cancer.

References

1. Elbashir, S. M., Harborth, J., Lendeckel, W., Yalcin, A., Weber, K., and Tuschl, T. (2001) Duplexes of 21-nucleotide RNAs mediate RNA interference in cultured mammalian cells. *Nature* **411**, 494–498.
2. Paddison, P. J., Caudy, A. A., Bernstein, E., Hannon, G. J., and Conklin, D. S. (2002) Short hairpin RNAs (shRNAs) induce sequence-specific silencing in mammalian cells. *Genes Dev* **16**, 948–958.
3. Brummelkamp, T. R., Bernards, R., and Agami, R. (2002) A system for stable expression of short interfering RNAs in mammalian cells. *Science* **296**, 550–553.
4. Paul, C. P., Good, P. D., Winer, I., and Engelke, D. R. (2002) Effective expression of small interfering RNA in human cells. *Nat Biotechnol* **20**, 505–508.
5. Sui, G., Soohoo, C., Affarel, B., Gay, F., Shi, Y., and Forrester, W. C. (2002) A DNA vector-based RNAi technology to suppress gene expression in mammalian cells. *Proc Natl Acad Sci USA* **99**, 5515–5520.
6. Miyagishi, M., and Taira, K. (2002) U6 promoter-driven siRNAs with four uridine 3′ overhangs efficiently suppress targeted gene expression in mammalian cells. *Nat Biotechnol* **20**, 497–500.
7. Lee, N. S., Dohjima, T., Bauer, G., et al. (2002) Expression of small interfering RNAs targeted against HIV-1 rev transcripts in human cells. *Nat Biotechnol* **20**, 500–505.
8. Berlivet, S., Guiraud, V., Houlard, M., and Gerard, M. (2007) pHYPER, a shRNA vector for high-efficiency RNA interference

in embryonic stem cells. *Biotechniques* **42**, 738–743.
9. Houlard, M., Berlivet, S., Probst, A. V., et al. (2006) CAF-1 is essential for heterochromatin organization in pluripotent embryonic cells. *PLoS Genet* **2**, e181.
10. Quivy, J. P., Roche, D., Kirschner, D., Tagami, H., Nakatani, Y., and Almouzni, G. (2004) A CAF-1 dependent pool of HP1 during heterochromatin duplication. *Embo J* **23**, 3516–3526.

Chapter 8

Regulation and/or Repression of Cholinergic Differentiation of Murine Embryonic Stem Cells Using RNAi Directed Against Transcription Factor L3/Lhx8

Takayuki Manabe and Akio Wanaka

Abstract

Techniques for controlling the expression of a specific gene in embryonic stem cells are effective and important for clarifying the functions of the gene. Regarding differentiation of cells into nervous system components, these techniques would play key roles in elucidating, not only the differentiation mechanisms of neuronal and glial cells, but also how neuronal phenotypes are determined. In this chapter, we describe an RNA interference method for suppressing cholinergic differentiation in murine embryonic stem cells by knockdown of expression of the transcription factor L3/Lhx8, a Lim homeobox gene family protein. This method will greatly facilitate functional analyses of the factors involved in neuronal differentiation and regeneration and contribute to cell transplantation studies.

Key words: Lim homeobox, L3/Lhx8, acetylcholine, embryonic stem cell, RNAi.

1. Introduction

To date, the functions of a specific gene at the individual animal level have been identified by creating knockout (KO) animals for the target gene and analyzing the resulting phenotypes. However, the KO method requires a lot of labor and time and is not practicable in analyzing many genes simultaneously. Therefore, more efficient methods that affect the control of gene functions in individual animals or cells than the KO method are required. RNA interference (RNAi) is a mechanism in which the expression of

B. Zhang, E.J. Stellwag (eds.), *RNAi and MicroRNA-Mediated Gene Regulation in Stem Cells*, Methods in Molecular Biology 650,
DOI 10.1007/978-1-60761-769-3_8, © Springer Science+Business Media, LLC 2010

a target gene is reduced when a double-stranded RNA (dsRNA) corresponding to part of the mRNA of the target gene is introduced into a cell. In 1998, the induction of a dsRNA for expression of the same gene as a transgene with repressor activity was discovered in *Caenorhabditis elegans* (1). In mammals, the effects of RNAi were reported for the first time following injection of dsRNA into early murine embryos (2). However, the effects were limited to the early embryos and disappeared by birth (2). Given the diminished effects, it was hypothesized that the introduced dsRNA failed to sustain activity because of dilution, and that a sufficiently high concentration could not be maintained to achieve RNAi. Recently, we have been able to obtain a sustained repressor effect following stable transfection of vector-based dsRNA into cells.

L3/Lhx8 is a recently identified member of the LIM homeobox gene family (3). In the brain, L3/Lhx8 is selectively expressed in the medial ganglionic eminence (3). We found that *L3/Lhx8*-KO mice lost a significant number of basal forebrain cholinergic neurons, while an L3/Lhx8-knockdown neuro2A neuroblastoma cell line specifically lacked the cholinergic phenotype (4, 5). Furthermore, we were able to control cholinergic differentiation by suppressing the expression of L3/Lhx8 by RNAi in murine embryonic stem (ES) cells (6, 7), as described in this chapter.

Murine ES cells contribute to the formation of all tissues and represent an ideal system for examining differentiation mechanisms in vitro. Oct3/4, which plays a critical role in maintaining ES cell pluripotency (8), can promote neuroectoderm formation by ES cells with a higher level of expression and subsequent neural differentiation (9). Oct3/4-expressing ES cells (EB5) allow us to achieve effective cholinergic differentiation of ES cells. Therefore, we used this ES cell line in our protocol.

We used a vector-based RNAi approach in this protocol to produce a short hairpin dsRNA intracellularly from a DNA template under the control of the histone H1.2 promoter (5–7). The H1.2 promoter in this vector is an engineered inducible promoter containing a tetracycline (Te) operator (TetO1). Stable clones generated using this vector express green fluorescence protein (GFP) in a Te-independent manner under the control of the cytomegalovirus (CMV) promoter.

Experimental conditions described in this protocol effectively induced neuronal (25.6 ± 2.2%) and astroglial (55.7 ± 6.9%) differentiation of ES cells (7). Differentiation into oligodendrocytes and microglial cells was observed only infrequently (7).

2. Materials

2.1. Cell Culture

1. Phosphate-buffered saline (PBS): Prepare stock solutions A (0.2 M Na_2HPO_3) and B (0.2 M NaH_2PO_3) and autoclave before storage at 4°C. Prepare a working solution by mixing 40.5 mL of solution A, 9.5 mL of solution B, and 9.9 g of NaCl and make up to 1 L with Milli-Q water, followed by autoclaving and storage at 4°C.
2. Gelatin solution: Prepare 0.1% gelatin in PBS, and autoclave before storage at 4°C. Coat individual culture dishes with the gelatin solution at 37°C for 1 h in CO_2 incubator (4 mL/10 cm dish). Remove the gelatin solution by pipetting and wash the dishes three times with PBS (5 mL) in clean bench. After drying, the plates can be used for cell culture.
3. Growth medium: Dulbecco's modified Eagle's medium (DMEM; Nacalai Tesque Japan, Kyoto, Japan) supplemented with 10% fetal bovine serum (FBS; Stem Cell Sciences, Kobe, Japan) (*see* **Note 1**), 2 mM glutamine, 0.1 mM nonessential amino acids, 1 mM pyruvate, 0.1 mM 2-mercaptoethanol (2-ME), 2000 U/mL LIF (Gibco/BRL, Bethesda, MD), 20 mg/mL blasticidin S (*see* **Note 2**), 100 U/mL penicillin, and 10 μg/mL streptomycin.
4. TrypsinEDTA solution (0.25%) from Gibco/BRL. Dilute this solution 1:10 with PBS immediately before use.
5. Cell Banker storage solution from Juji Field Inc. (Tokyo, Japan).

2.2. Plasmids and Transfection

1. pRNATin-H1.2/Hygro vector from GenScript Corporation (Piscataway, NJ). pcDNA3.1(+) and pcDNA6/TR vectors from Invitrogen (Carlsbad, CA).
2. Oligonucleotides (L3/Lhx8 siRNA) synthesized by Takara Bio Inc. (Shiga, Japan). The sequences of the oligonucleotides are as follows: 5′-*GGATCC*CGTGCCAGCATAAGTCATTCACCTTGATATCCGGGTGAATGACTTATGCTGGCA**TTTTTT**CCAA*AAGCTT*-3′; 5′-*AAGCTT*TTGG**AAAAAA**TGCCAGCATAAGTCATTCACCCGGATATCAAGGTGAATGACTTATGCTGGCACG*GGATCC*-3′. The underlined, bold, and italicized letters denote the hairpin loop, terminal signal, and target sites for the restriction enzymes *BamHI* (GGATCC) and *HindIII* (AAGCTT) (*see* **Note 3**).
3. Targefect F-1 from Targeting Systems (Santee, CA).
4. Opti-MEM from Gibco/BRL.

2.3. Knockdown of L3/Lhx8

1. Sphere formation medium: DMEM supplemented with 10% knockout serum replacement (KSR; Gibco/BRL), 2 mM glutamine, 0.1 mM nonessential amino acids, 1 mM pyruvate, and 0.1 mM 2-ME.
2. Differentiation medium: Sphere formation medium containing 0.1 μM retinoic acid (RA).
3. Tetracycline hydrochloride (Te; Wako Pure Chemical Industries, Osaka, Japan) stock solution: 50 μg/mL (1,000×) Te in distilled water (*see* **Note 4**). Sterilize the stock solution using a 0.22-μm filter. The stock solution can be stored at –20°C.
4. P/L/G coating solution: Mix poly-L-ornithine (Sigma, St. Louis, MO; 0.002% in PBS), laminin (Invitrogen; 5 μg/mL in PBS), and gelatin (*see* **Section 2.1 and step 2**; 0.1% in PBS) at a ratio of 1:1:1 immediately before use. Coat individual culture dishes with the P/L/G coating solution at 37°C for 1 h. Remove the P/L/G solution and wash the dishes three times with PBS (5 mL) in clean bench. After drying, the dishes can be used for cell culture.
5. Trizol solution form Life Technologies Japan (Tokyo, Japan).

2.4. Immunocytochemistry

1. PBS: *see* **Section 2.1 and step1**.
2. Fixation solution: 4% paraformaldehyde in PBS. This solution can be stored at 4°C in the dark for 1 week. Carefully heat this solution for complete dissolution by using a stirring hot-plate in a vented hood.
3. Blocking solution: 3% BSA and 0.5% Triton X-100 in PBS (make up immediately before use).
4. Antibody dilution buffer: Blocking solution.
5. Primary antibodies: Anticholine acetyltransferase (ChAT; mouse monoclonal; Chemicon, Temecula, CA) and anti-GFP (rabbit polyclonal; Molecular Probes Inc., Eugene, OR).
6. Secondary antibodies: Alexa 546-conjugated antimouse IgG and Alexa 488-conjugated antirabbit IgG (both from Molecular Probes Inc.).
7. Mounting solution: Antifade (Molecular Probes).

2.5. Cell Death Assay

1. Terminal deoxynucleotidyl transferase-mediated dUTP nick end labeling (TUNEL): In Situ Cell Death Detection Kit (Roche Diagnostics K.K., Tokyo, Japan).

3. Methods

3.1. Cell Culture

1. Mouse ES cells (EB5) are maintained on gelatin-coated dishes in growth medium. The cells are incubated at 37°C in a humidified atmosphere of 95% air/5% CO_2 and subcultured into fresh culture vessels when the growth reaches 7090% confluence (i.e., every 23 days). The growth medium is exchanged for fresh medium every 2 days. If necessary, the cells are stored in Cell Banker storage solution in the vapor phase of liquid nitrogen (*see* **Note 5**).

3.2. Plasmids and Transfection

All the experiments are carried out in an aseptic manner using typical clean bench procedures for cell culture.

1. ES cells are cultured on gelatin-coated 10-cm dishes in growth medium (*see* **Section 3.1 and step1**).
2. After 2 days, the culture dishes are taken out and the medium is exchanged for Opti-MEM. The dishes are then returned to the incubator.
3. pcDNA6/TR (1 μg) is mixed with Opti-MEM (100 μL) in a 1.5-mL tube (solution C). Next, 3 μL of Targefect F-1 is mixed with the medium in the 1.5-mL tube (solution D).
4. Solution C is mixed with solution D, and the mixed solution is allowed to incubate for 20 min at room temperature.
5. After the 20-min reaction, the mixed solution is added to the cells and incubated for 6 h at 37°C in CO_2 incubator.
6. The cells are cultured on new gelatin-coated 10-cm dishes in growth medium containing 200 μg/mL of G418 (geneticin) (*see* **Note 6**).
7. The medium is exchanged for fresh medium containing G418 every 2 days.
8. After 10 days, 1020 clones will be obtained. These cells are collected by using the trypsinEDTA solution and cultured on new gelatin-coated 10-cm dishes (*see* **Note** 7).
9. After 2 days, the culture dishes are taken out and the medium is exchanged for Opti-MEM. The dishes are then returned to the incubator.
10. pRNATin-H1.2-L3/Lhx8-siRNA (1 μg) is mixed with Opti-MEM (100 μL) in a 1.5-mL tube (solution E). Targefect F-1 (3 μL) is mixed with the medium in a 1.5-mL tube (solution F). The same operation is carried out for the pRNATin-H1.2 vector (Mock) as described below.

11. Solution E is mixed with solution F, and the mixed solution is allowed to incubate for 20 min at room temperature.
12. After the 20-min reaction, the mixed solution is added to the cells and incubated for 6 h at 37°C in CO_2 incubator.
13. The cells are passaged on new gelatin-coated 10-cm dishes and cultured in growth medium containing 30 μg/mL of hygromycin.
14. The medium is exchanged for fresh medium containing hygromycin every 2 days.
15. About 12 clones are picked, and these clones are plated and cultivated individually. These procedures and the following procedures are carried out in medium containing 15 μg/mL of hygromycin and 100 μg/mL of G418.

3.3. Knockdown of L3/Lhx8

1. Individual clones are plated on 10-cm bacterial grade dishes in sphere formation medium with or without (control) 50 ng/mL of Te (*see* **Section 2.3 and step 3**) for 4 days.
2. The cells are collected by centrifugation at 100*g* for 5 min, washed with PBS, and treated with trypsin for 5 min at 37°C.
3. The trypsinized cells are dissociated by pipetting, centrifuged after dilution with an equal volume of differentiation medium, and resuspended with differentiation medium with or without Te (1.5×10^5 cells/1.5 mL/35-mm dish).
4. Cells are plated on P/L/G-coated 35-mm dishes at 1.5×10^5 cells/dish in differentiation medium with or without Te. Following incubation for 1 h, the differentiation medium is replaced with fresh medium containing 0.1 μM RA. The medium containing 0.1 μM RA is changed every 3 days, and the cells are allowed to differentiate for 14 days (*see* **Note 8**).
5. After 14 days, mRNA is isolated from the cells by a generalized Trizol extraction method, and the effects of the knockdown are examined (*see* **Note 9**). The most promising clone is used for experiments as follows.
6. Steps 13 are carried out using the most promising clone.
7. Cells are plated on P/L/G-coated glass-bottom dishes (35 mm) at 1.5×10^5 cells/dish in differentiation medium with or without Te. Following incubation for 1 h, the differentiation medium is replaced with fresh medium containing 0.1 μM RA. The medium containing 0.1 μM RA is changed every 3 days, and the cells are allowed to differentiate for 14 days.

3.4. Immunocytochemistry

The following procedures are carried out carefully to avoid dislodging the cells from the dishes, and promptly to avoid drying the cells.

1. The differentiated cells obtained in **Section 3.3 and step7** are rinsed with PBS.
2. Fixation solution is then added for 10 min at room temperature (*see* **Note 10**).
3. The fixation solution is discarded and the cells are washed 3 times with PBS (1.5 mL) for 5 min each at room temperature.
4. The cells are blocked by incubation in 1.5-mL blocking solution for 2 h at room temperature (*see* **Note 10**).
5. The blocking solution is removed and replaced with anti-GFP (1:5000) and anti-ChAT (1:1000) antibodies in antibody dilution buffer for 2 h at room temperature (*see* **Note 10**).
6. The primary antibodies are removed and the cells are washed 3 times with PBS (1.5 mL) for 5 min each at room temperature.
7. The secondary antibodies are prepared at 1:1000 in antibody dilution buffer and incubated with the cells under dark conditions (in aluminum foil) for 1 h at room temperature. The following operations are carried out under dimmed light.
8. The secondary antibodies are discarded and the cells are washed 3 times with PBS (1.5 mL) for 20 min each.
9. After complete removal of the PBS, mounting solution is added and the cells are covered with a coverslip to avoid introduction of air. Excess mounting solution is absorbed later.
10. The cells are observed under a fluorescence microscope. The total numbers of cells (100–150 cells in 4 different fields at 100× magnification) are counted using the GFP immunofluorescence (green emission) to identify the cells in a single experiment. Cells positive for ChAT (red emission) are quantified in the same fields at the same magnification. The counting experiments are performed in triplicate for each marker. The results are expressed as the mean ± SE and analyzed statistically by Student's *t*-test.

3.5. Cell Death Assay

To confirm that biased cell death is not caused during a series of these experiments, a cell death assay is carried out.

1. The differentiated cells obtained in **Section 3.3 and step7** are processed for TUNEL staining using an In Situ Cell

Death Detection Kit (Roche) according to the manufacturer's protocol.

2. The TUNEL-positive cells are observed under a microscope. The total numbers of cells (100–150 cells in 4 different fields at 100× magnification) are counted. Cells positive for TUNEL staining are quantified in the same fields at the same magnification. The counting experiments are performed in triplicate for each marker. The results are expressed as the mean ± SE and analyzed statistically by Student's *t*-test.

4. Notes

1. This FBS is high-grade FBS for ES cells.
2. ES cells (EB5), a kind gift from Dr. Hitoshi Niwa (RIKEN), carry the blasticidin S-resistance selection marker gene driven by the Oct3/4 promoter. The EB5 cell line was derived from E14tg2a ES cells (10) and generated by targeted integration of the Oct-3/4-IRES-BSD-pA vector (8) into the *Oct-3/4* allele.
3. Annealed oligonucleotides were cloned into the *BamHI-HindIII* sites of the pRNATin-H1.2/Hygro vector. The siRNA was designed using the Web-based siRNA design program from the GenScript Corporation Web page (http://www.genscript.com/rnai.html). The sequence used avoided the conserved LIM homeobox domains and only produced a specific hit for *L3/Lhx8* in the GenBank database.
4. Tetracycline that is not the hydrochloride salt is solubilized in ethanol.
5. The ES cells are stocked routinely in the vapor phase of liquid nitrogen.
6. G418 should not be used at concentrations above 200 μg/mL for ES cells, unlike other cell lines.
7. Clones do not need to be made at this stage because the most promising clone, i.e., the one with high knockdown activity is selected by the subsequent procedure. Since high-expression clones do not necessarily have high knockdown activity, clones should not be selected here.
8. The experimental conditions (concentrations, times of addition, periods of addition, etc.) for the optimum Te and RA treatment regimen must be independently established for each individual target gene.

9. If an antibody for the target protein exists, the protein expression level should be examined by immunoblotting.
10. An overnight incubation at 4°C is acceptable.

Acknowledgments

We thank Dr. H. Niwa (RIKEN) for providing the murine ES cells. T.M. was supported by a grant from The Ichiro Kanehara Foundation and a Grant-in-Aid for Scientific Research on Priority Areas, Advanced Brain Science Project, from the Ministry of Education, Culture, Sports, Science and Technology, Japan.

References

1. Fire, A., Xu, S., Montgomery, M. K., Kostas, S. A., Driver, S. E., and Mello ,C. C. (1998) Potent and specific genetic interference by double-stranded RNA in *Caenorhabditis elegans. Nature* **391,** 806–811.
2. Wianny, F., and Zernikca-Goetz, M. (2000) Specific interference with gene function by double-stranded RNA in early mouse development. *Nature Cell Biol* **2,** 70–75.
3. Matsumoto, K., Tanaka, T., Furuyama, T., Kashihara, Y., Mori, T., Ishii, N., Kitanaka, J., Takemura, M., Tohyama, M., and Wanaka, A. (1996) L3, a novel murine LIM-homeodomain transcription factor expressed in the ventral telencephalon and the mesenchyme surrounding the oral cavity. *Neurosci Lett* **204,** 113–116.
4. Mori, T., Yuxing, Z., Takaki, H., Takeuchi, M., Iseki, K., Hagino, S., Kitanaka, J., Takemura, M., Misawa, H., Ikawa, M., Okabe, M., and Wanaka, A. (2004) The LIM homeobox gene, L3/Lhx8, is necessary for proper development of basal forebrain cholinergic neurons. *Eur. J. Neurosci.* **19,** 3129–3141.
5. Manabe, T., Tatsumi, K., Inoue, M., Matsuyoshi, H., Makinodan, M., Yokoyama, S., and Wanaka, A. (2005) L3/Lhx8 is involved in the determination of cholinergic or GABAergic cell fate. *J. Neurochem.* **94,** 723–730.
6. Manabe, T., Tatsumi, K., Inoue, M., Makinodan, M., Yamauchi, T., Makinodan, E., Yokoyama, S., Sakumura, R., and Wanaka, A. (2007) L3/Lhx8 is a pivotal factor for cholinergic differentiation of murine embryonic stem cells. *Cell Death Differ.* **14,** 1080–1085.
7. Manabe, T., Tatsumi, K., Inoue, M., Matsuyoshi, H., Makinodan, M., Yamauchi, T., Makinodan, E., Yokoyama, S., Sakumura, R., Okuda, H., and Wanaka, A. (2008) Knockdown of the L3/Lhx8 gene suppresses cholinergic differentiation of murine embryonic stem cell-derived spheres. *Int J Dev Neurosci.* **26,** 1249–252.
8. Niwa, H., Miyazaki, J., and Smith, AG. (2000) Quantitative expression of Oct-3/4 defines differentiation, dedifferentiation or self-renewal of ES cells. *Nat Genet.* **24,** 372–376.
9. Shimozaki, K., Nakashima, K., Niwa, H., and Taga, T. (2003) Involvement of Oct3/4 in the enhancement of neuronal differentiation of ES cells in neurogenesis-inducing cultures. *Development* 130, 2505–2512.
10. Hooper, M., Hardy, K., Handyside, A., Hunter, S., and Monk, M. (1987) HPRT-deficient (Lesch-Nyhan) mouse embryos derived from germline colonization by cultured cells. *Nature* **326,** 1292–295.

Chapter 9

Silencing of Rho-GDIγ by RNAi Promotes the Differentiation of Neural Stem Cells

Jiao Wang, Wei Lu, and Tieqiao Wen

Abstract

RNA interference (RNAi) technology is one of the main means in the study of stem cell differentiation. This study describes Rho-GDIγ function during the differentiation of neural stem cells by using RNAi. Rho-GDIγ belongs to the Rho-GDI protein family, which is expressed at high level throughout the brain. Although it exists in neuronal population, its physiological function is poorly understood. By using RNAi technology to downregulate expression of Rho-GDIγ, we found distinct morphological changes in neural stem cell line C17.2. More important, RT-PCR confirmed that RNAi-mediated downregulation of Rho-GDIγ decreased expression of Rho-GDIγ-regulated genes RhoA and slightly increased expression of Rac1. Further, immunochemical staining indicated that downregulation of Rho-GDIγ increased the tendency of C17.2 cells to differentiate. These data strongly suggest that Rho-GDIγ plays a key role in the differentiation of neural stem cells.

Key words: RNAi, cells culture, differentiation, regulation.

1. Introduction

The phenomenon of RNA interference (RNAi) was first discovered in *Caenorhabditis elegans* by Andrew Fire et al. (1). They found a potent gene-silencing effect after injecting double-stranded RNA into *C. elegans*. From then on, RNAi technology has revealed a new regulatory paradigm in biology. It is one of the principal means within living cells that helps to specifically down-regulate interested genes which are active and has been used successfully in many studies of neural stem cell differentiation. For example, RNAi has been used in neuronal cell lines (2), neural stem cells (3), primary mammalian neurons (4), astrocytes (5), and Schwann cells (6).

B. Zhang, E.J. Stellwag (eds.), *RNAi and MicroRNA-Mediated Gene Regulation in Stem Cells*, Methods in Molecular Biology 650,
DOI 10.1007/978-1-60761-769-3_9,

Neural stem cells (NSCs) are self-renewing and have the potential to differentiate into neurons, astrocytes, and oligodendrocytes (7, 8). Their properties of stable self-renewal and multipotential differentiation make NSCs an attractive and presumably unlimited donor source for cell replacement therapy to treat neurodegenerative disease (9). Rho proteins belong to the small GTP-binding protein (G protein) superfamily, which includes the Rho, Rac, and Cdc42 subfamilies. Rho family members facilitate reorganization of the actin cytoskeleton, regulating diverse cellular processes including cellular morphology, aggregation, and adhesion. Rho proteins exist in two interconvertible forms: a GDP-bound inactive form and a GTP-bound active form. Rho-GDIγ is a member of the Rho family and plays a key role in modulating the activity of GTPases. It acts as a molecular switch in a dual manner in signal transduction pathways that regulate a multitude of biological processes including cell proliferation, apoptosis, differentiation, cytoskeletal reorganization, and membrane trafficking (10). Rho-GDIγ gene is now designated ARHGDIG by HUGO, which observed high levels of expression in the entire brain, with regional variations (11). Three Rho-GDIs, α, β, and γ, have been identified. Rho-GDIα was cloned from brain and is ubiquitously expressed. Rho-GDIβ is preferentially expressed in hematopoietic cells. Rho-GDIγ was cloned from a whole embryo library and is highly expressed in brain and pancreas (12). Rho-GDI proteins are one of the central regulators of the RhoA/Rac family of small GTPases.

Previous microarray studies demonstrated that >10,000 genes are expressed during differentiation of neural stem cells (13). This study uses RNAi technology (the detailed procedure of experiment is shown in **Fig. 9.1**) to analyze the role of Rho-GDIγ in differentiation of neural stem cells (8).

2. Materials

2.1. Cells Culture and Transfection

1. High-glucose Dulbecco's Modified Eagle's Medium (DMEM) (Invitrogen) supplemented with 10% fetal bovine serum (Invitrogen) and 5% heat-inactivated horse serum (Invitrogen); storage at 4 degrees.
2. Solution of trypsin (0.25%) and ethylenediamine tetraacetic acid (EDTA) (1 mM) from Gibco/BRL.
3. Phosphate-buffered saline (PBS) Wash Buffer (1X): 140 mM NaCl, 2.7 mM KCl, 10 mM Na_2HPO_4, 1.8 mM KH_2PO_4dissolved in distilled autoclaved water. The pH has been adjusted to 7.4 using hydrochloric acid. Autoclave before storage at room temperature.

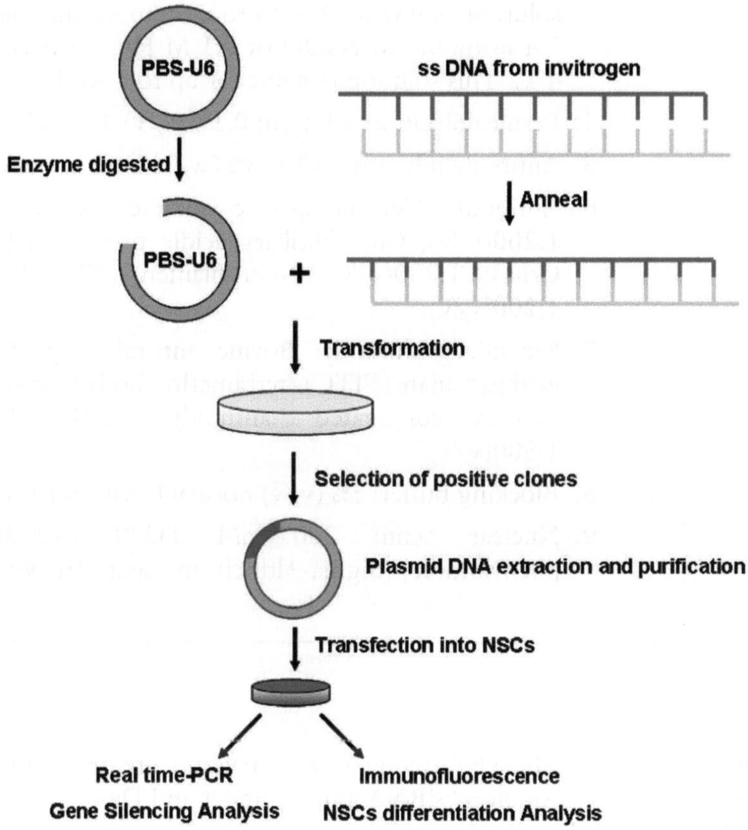

Fig. 9.1. Illustration of the main experimental procedure of RNAi technology.

4. Plasmid DNA extraction kit: PureLink^{TM} Quick Plasmid Miniprep Kit (Invitrogen).
5. Restriction endonucleases: *Apa I*, *HindIII*, and *EcoRI*.
6. Lipofectamine 2000 from Invitrogen is used as plasmid transfection reagent.

2.2. Real-Time PCR

1. Chloroform/isopropanol.
2. 75% Ethanol (cold)/DEPC (diethyl pyrocarbonate)-treated water (Sigma).
3. Trizol Isolation Reagent (TaKaRa).
4. PCR tubes, microcentrifuge tubes, pipette tips. (*See* **Note 1**).

2.3. Immunocytochemical Staining

1. Phosphate-buffered saline (PBS): the same as in **Section 2.1** and **step 3**.
2. 35-mm Glass bottom culture dishes: Fluoro Dish (Corning).
3. Paraformaldehyde fix solution: Add 4 g paraformaldehyde to 100 mL PBS + azide. Heat to 70°C in a fume hood or in a 56°C water bath just until the paraformaldehyde goes into

solution. Allow to cool to room temperature, adjust to pH 7.4 using 0.1 M NaOH or 0.1 M HCl, as needed. Store at 4°C. This solution is stable for up to 1 week.

4. Permeabilization solution: 0.1% (v/v) Triton X-100 in PBS.
5. Antibody dilution buffer: 3% (w/v) BSA in PBS.
6. Antibody: Neuron specific enolase (NSE) (Chemicon): 1:200(v/v); Glial fibrillary acidic protein (GFAP) (Santa Cruz): 1:200(v/v); Neurofilamen (NF) (Santa Cruz): 1:200(v/v).
7. Secondary antibody: Bovine anti-rabbit/goat fluorescein isothiocyanate (FITC)/tetramethylrhodamineisothiocyanate (TRITC)-conjugated antibody. (BD Pharmingen): 1:500(v/v).
8. Blocking buffer: 5% (v/v) normal bovine serum in PBS.
9. Nuclear stain: 300 nM DAPI (4,6-diamidino-2-phenylindole) (Sigma-Aldrich) in water (*See* **Note 2**).

3. Methods

3.1. Design and Construction of RNAi-Expressing Plasmids

1. The RNAi sequence and structure are determined using the Ambion "siRNA Target Finder and Design Tool."
2. siRNA target sites are chosen by scanning an mRNA sequence for AA dinucleotides, recording the 19 nucleotides immediately downstream of the AA, and then comparing the potential siRNA target sequences with an appropriate genome database to ensure the absence of significant homology to other genes. When expressed in vivo, the insert forms a stem-loop hairpin structure.
3. The RNAi structures are as follows: (*See* **Note 3**)

	stem1	loop	stem2
R1	5′-ACCAGGTGTTTGTCCTGAA	TTCAAGAGA	TTCAGGACAAACACCT GGTTT TTTT-3′ (53 bp)
	5′-AATTAAAAAAACCAGGTGTTTG TCCTGAA	TCTCTTGAA	TTCAGGACAAACACCT GGT GGCC-3′ (61 bp).
R2	5′-ATGTCTGCATCACACCTAT	TTCAAGAGA	ATAGGTGTGATGCAGA CATTT TTTT-3′ (53 bp)
	5′-AATTAAAAAAATGTCTGCATCA CACCTAT	TCTCTTGAA	ATAGGTGTGATGCAGA CAT GGCC-3′ (61 bp).
Ck:	5′-GTGCTACAGTACTGGTCTA	TTCAAGAGA	TAGACCAGTACTGTAG CACTT TTTT-3′ (53 bp)
	5′-AATTAAAAAAGTGCTACAGTAC TGGTCTA	TCTCTTGAA	TAGACCAGTACTGTAGCA CG GCC-3′ (61 bp).

4. The linearized pBS/U6 siRNA Expression Vector is digested with *Apa I* and *EcoRI* and gel purified, followed by rapid ligation of siRNA insert oligonucleotides into the Pol III expression domain through a one-step method.
5. Positive clones containing the siRNA insert (**Fig. 9.2**) are selected by their resistance to digestion with HindIII. Positive clones are confirmed by sequencing the cloning site using a T3 promoter primer. The constructs are named pBS/U6-Ck, pBS/U6-R1, pBS/U6-R2, containing Ck, R1, or R2 oligonucleotides, respectively (*See* **Note 4**).

3.2. Cells Culture, Transfection, and Cell Morphology

1. Neural stem cell line C17.2 are grown in a flask at 37°C in a 1:1 mixture of Dulbecco's modified Eagle's medium and Ham's F-12 medium (Invitrogen) supplemented with 5% fetal bovine serum (Invitrogen) and 5% heat-inactivated horse serum (Invitrogen) in a 5% CO_2-humidified chamber.
2. Cells are split at a density of 2×10^5 cells per 60-mm dish.
3. Transfection is carried out 4 h later using the lipofectamine 2000 method (Invitrogen). (*See* **Note 5**).
4. RNAi-expressing plasmids are constructed to determine the effect of downregulating Rho-GDIγ in neural stem cells. The structures of the RNAi-coding insert and plasmid construction are shown in **Fig. 9.2**.

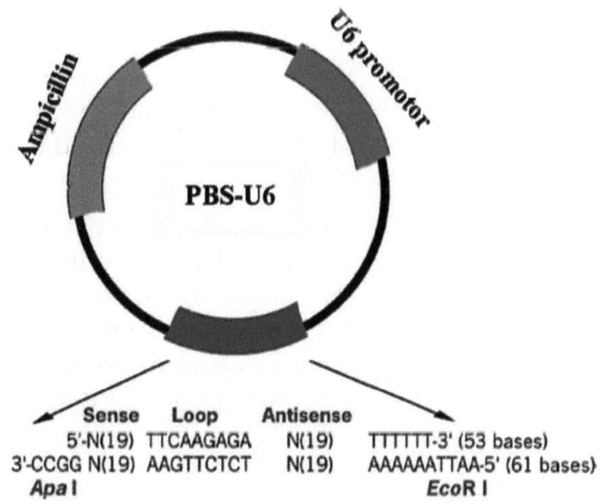

Fig. 9.2. Schematic diagram of RNAi constructs and RNAi insert sequence. The linearized pBS/U6 siRNA expression vector is digested with Apa I and EcoRI and gel purified, followed by rapid ligation of siRNA insert oligonucleotides into the Pol III expression domain through a one-step method.

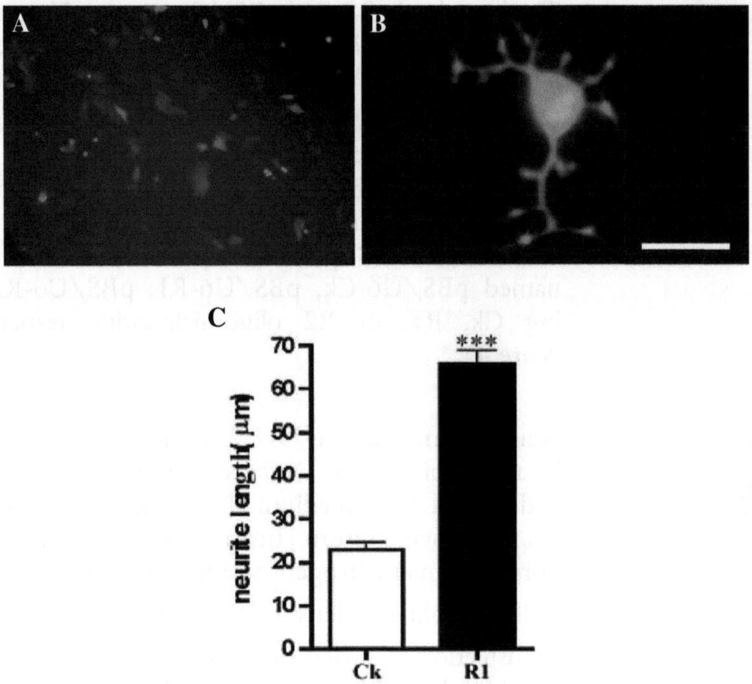

Fig. 9.3. Effects of Rho-GDIγ-targeted RNAi on differentiation of neural stem cells. **(A)** Green fluorescence correlates with the efficiency of co-transfection with pC1-EGFP. **(B)** Cells co-transfected with pC1-EGFP and pBS/U6-R1 appear in the differentiated status. Scale bar equals 50 μm. **(C)** Neurite length in cells expressing Rho-GDIγ-targeted RNAi (pBS/U6-R1; black bar) or control RNAi (pBS/U6-Ck; white bar). (**$p < 0.001$).

5. RNAi-coding plasmids are co-transfected with pC1-EGFP (expressing enhanced green fluorescent protein, EGFP) into C17.2 cells (**Fig. 9.3A**).
6. Photomicrographs are taken with an inverted phase contrast microscope (Nikon, Japan). The results showed that the constructs expressing R1 and R2 RNAi molecules (pBS/U6-R1 and pBS/U6-R2) suppressed expression of Rho-GDIγ, while the control plasmid (pBS/U6-Ck) did not.
7. Five days after the cells are co-transfected with pC1-EGFP and pBS/U6-R1, cells are observed under an inverted microscope to monitor cell morphology and differentiation. Control cells expressing pC1-EGFP and pBS/U6-Ck are slightly dispersed with rather short neurites. In contrast, cells expressing pC1-EGFP and pBS/U6-R1 have notably longer neurites and showed significant morphological changes (**Fig. 9.3B**).
8. Cell patterns are examined more quantitatively by measuring neurite length and number in six randomly selected microscopic fields. The results show statistically significant differences from control for neurite length (**$p < 0.001$) (**Fig. 9.3C**) in cells expressing Rho-GDIγ-targeted RNAi.

3.3. Reverse Transcription (RT) PCR and Real-Time PCR

1. Total RNA is isolated from cultured cells using total RNA extraction kit after incubation for 2 days.
2. First-strand cDNA synthesis is carried out using oligo (dT) primer and AMV reverse transcriptase according to the manufacturer's instructions (*See* **Note 6**).
3. Primer sets are synthesized for gene amplification of GAPDH, Rho-GDIγ, RhoA, and Rac1.
4. The expression level is normalized to GAPDH (internal control).
5. For Real-Time PCR, SYBR Green I is used with the cycles set to 40–45 and C_t set to 0.25. **Fig. 9.4A** (*See* **Note 7**).
6. The silence efficiency was calculated in **Fig. 9.4B**: Rho-GDIγ (−6.87), RhoA (−23.70), and Rac1 (1.10).
7. Thus, expression of Rho-GDIγ and Rho-GDIγ-regulated genes except Rac1 was downregulated, while expression of Rac1 was slightly upregulated.

3.4. Immunocytochemical Staining

1. NSE, NF, and GFAP staining is performed according to nestin immunocytochemical staining as described by Meng et al (14). One week after the cells are co transfected with pC1-EGFP and RNAi constructs, cells are rinsed and immunochemical staining for neuronal marker genes using fluorescence-tagged secondary antibodies.

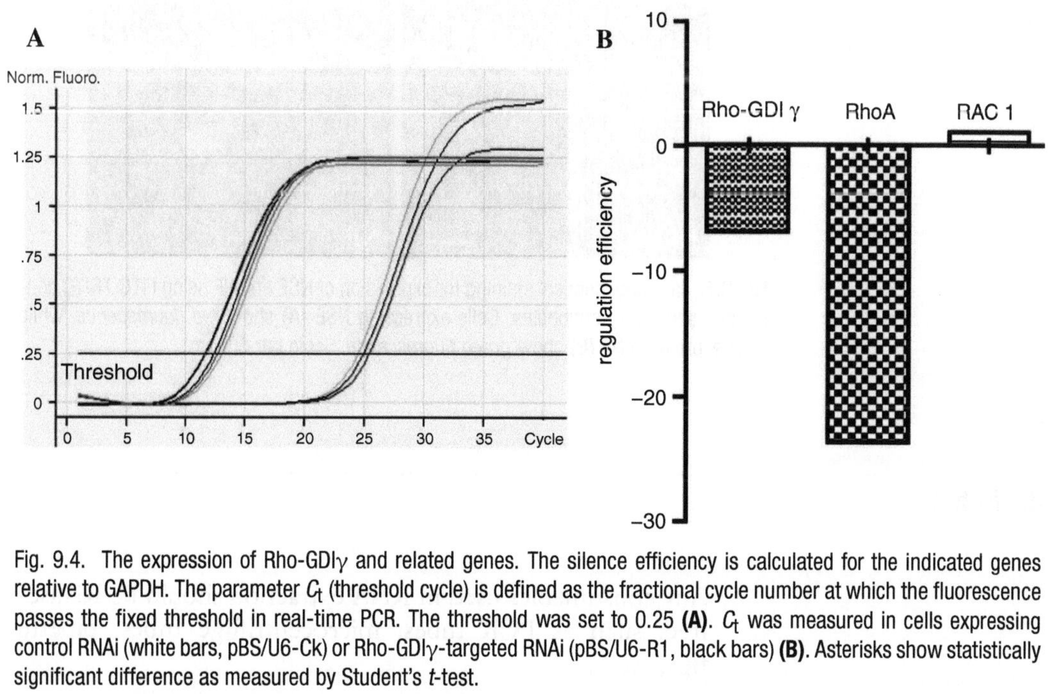

Fig. 9.4. The expression of Rho-GDIγ and related genes. The silence efficiency is calculated for the indicated genes relative to GAPDH. The parameter C_t (threshold cycle) is defined as the fractional cycle number at which the fluorescence passes the fixed threshold in real-time PCR. The threshold was set to 0.25 **(A)**. C_t was measured in cells expressing control RNAi (white bars, pBS/U6-Ck) or Rho-GDIγ-targeted RNAi (pBS/U6-R1, black bars) **(B)**. Asterisks show statistically significant difference as measured by Student's *t*-test.

2. Cells are fixed with 4% paraformaldehyde, permeabilized with 0.1% Triton X-100 in PBS three times, and blocked with 5% normal bovine serum for 30 min.
3. Cells are incubated with a 1:500 dilution of rabbit polyclonal anti-NSE antibody (1:500; Santa Cruz Biotechnology) and goat polyclonal anti-NF antibody (1:500; Santa Cruz Biotechnology) in blocking buffer for 1 h (*See* **Note 8**).
4. Cells are washed three times in PBS and incubated with a TRITC-conjugated anti-rabbit immunoglobulin antibody (1:200; Santa Cruz Biotechnology) and a tetramethylrhodamine isothiocyanate (FITC)-conjugated anti-goat antibody (1:200; Santa Cruz Biotechnology) in blocking buffer for 1 h.
5. Fluorescence is detected using a Nikon microscope. Cells expressing the control RNAi construct pBS/U6-Ck are regular in shape and did not express NSE or GFAP (Data not shown). In contrast, cells expressing Rho-GDIγ-targeted RNAi have long neurites and altered cell shape, and expressed NSE (**Fig. 9.5A**) and NF (**Fig. 9.5B**) but not GFAP (Data not shown).
6. This result suggests that downregulation of Rho-GDIγ promotes differentiation of neural stem cells and that Rho-GDIγ may be a negative regulator of neural differentiation.

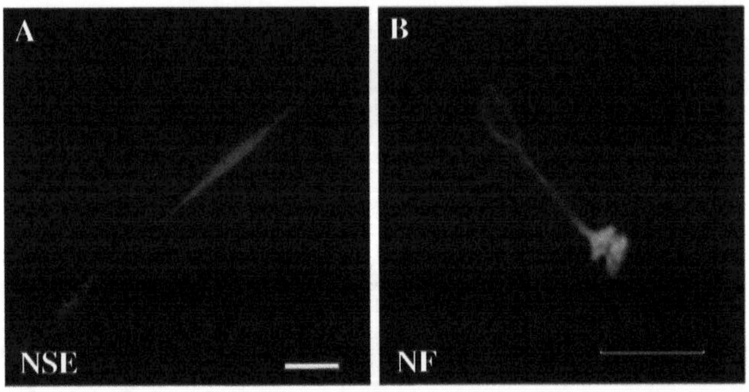

Fig. 9.5. Immunochemical staining for expression of NSE and NF using FITC/TRITC conjugated secondary antibodies. Cells expressing NSE **(A)** show *red fluorescence*, while cells expressing NF **(B)** show *green fluorescence*. Scale bar 50 μm.

4. Notes

1. Any consumables used in RNA extraction should be RNase-free, such as PCR tubes, microcentrifuge tubes, pipette tips, etc.

2. Water should have a resistivity of 18.2 MΩ-cm.
3. R1 was targeted to the middle of gene Rho-GDIγ, R2 was targeted to the 3′ end of Rho-GDIγ, and Ck was the control which used a scrambled version of R1.
4. To ensure the effectiveness of silence, positive clones containing the siRNA insert should be sequenced. It can be considered T3 promoter sequences as primer.
5. In view of toxicity of transfection reagent, lipofectamine 2000 dosage needs to be tried for different cell lines.
6. To avoid degradation of RNA, preferably, reverse transcription should be carried out immediately after RNA extraction. Otherwise, keep RNA solution in a −80°C refrigerator.
7. Before Real-Time PCR, conventional PCR should be done so that you can use the optimum template concentration during Real-Time PCR.
8. The antibody dose as well as holding time should be adjusted according to different cells. In addition, primary antibody can be used repeatedly.

Acknowledgments

We thank Prof. Snyder, E.Y for kindly donating C17.2 for this research. This work was supported by grants from the Major State Basic Research and Development Program of China (973 program, Grant No. 2006CB500702), the National Natural Science Foundation of China (Grant No. 30570590, 30770695), the Shanghai Commission of Science and Technology Basic Research Fund (Grant No. 03JC14030), the Shanghai Commission of Education Science and Technology Innovation Fund (Grant No. 08 ZZ41).

References

1. Fire, A., Xu, S., Montgomery, M. K., Kostas, S. A., Driver, S. E., and Mello, C. C. (1998) Potent and specific genetic interference by double-stranded RNA in *Caenorhabditis elegans. Nature* **391**, 806–811.
2. Gan, L., Anton, K. E., Masterson, B. A., Vincent, V. A., Ye, S., and Gonzalez-Zulueta, M. (2002) Specific interference with gene expression and gene function mediated by long dsRNA in neural cells. *J Neurosci Methods* **121**, 151–157.
3. Wood, M. J., Trulzsch, B., Abdelgany, A., and Beeson, D. (2003) Therapeutic gene silencing in the nervous system. *Hum Mol Genet* **12**(2), R279–284.
4. Krichevsky, A. M., and Kosik, K. S. (2002) RNAi functions in cultured mammalian neurons. *Proc Natl Acad Sci USA* **99**, 11926–11929.
5. Nicchia, G. P., Frigeri, A., Liuzzi, G. M., and Svelto, M. (2003) Inhibition of aquaporin-4 expression in astrocytes by RNAi determines alteration in cell morphology, growth, and water transport and induces changes in ischemia-related genes. *Faseb J* **17**, 1508–1510.

6. Higuchi, H., Yamashita, T., Yoshikawa, H., and Tohyama, M. (2003) Functional inhibition of the p75 receptor using a small interfering RNA. *Biochem Biophys Res Commun* **301**, 804–809.
7. Merkle, F. T., Mirzadeh, Z., and Alvarez-Buylla, A. (2007) Mosaic organization of neural stem cells in the adult brain, *Science (New York, NY)* **317**, 381–384.
8. Lu, W., Wang, J., and Wen, T. (2008) Down-regulation of Rho-GDI gamma promotes differentiation of neural stem cells. *Mol Cell Biochem* **311**, 233–240.
9. Newman, M. B., and Bakay, R. A. (2008) Therapeutic potentials of human embryonic stem cells in Parkinson's disease. *Neurotherapeutics* **5**, 237–251.
10. Bar-Sagi, D., and Hall, A. (2000) Ras and Rho GTPases: A family reunion. *Cell* **103**, 227–238.
11. Adra, C. N., Iyengar, A. R., Syed, F. A., Kanaan, I. N., Rilo, H. L., Yu, W., Kheraj, R., Lin, S. R., Horiuchi, T., Khan, S., Weremowicz, S., Lim, B., Morton, C. C., and Higgs, D. R. (1998) Human ARHGDIG, a GDP-dissociation inhibitor for Rho proteins: Genomic structure, sequence, expression analysis, and mapping to chromosome 16p13.3. *Genomics* **53**, 104–109.
12. Ferland, R. J., Li, X., Buhlmann, J. E., Bu, X., Walsh, C. A., and Lim, B. (2005) Characterization of Rho-GDIgamma and Rho-GDIalpha mRNA in the developing and mature brain with an analysis of mice with targeted deletions of Rho-GDIgamma, *Brain Res* **1054**, 9–21.
13. Wen, T., Gu, P., Minning, T. A., Wu, Q., Liu, M., Chen, F., Liu, H., and Huang, H. (2002) Microarray analysis of neural stem cell differentiation in the striatum of the fetal rat, *Cell Mol Neurobiol* **22**, 407–416.
14. Chalisova, N. I., Pennijajnen, V. P., Baskova, I. P., Zavalova, L. L., and Bazanova, A. V. (2003) The neurite-stimulating activity of components of the salivary gland secretion of the medicinal leech in cultures of sensory neurons, *Neurosci Behav Physiol* **33**, 411–414.

Chapter 10

RNAi Knockdown of Redox Signaling Protein Ape1 in the Differentiation of Mouse Embryonic Stem Cells

Gang-Ming Zou, Cynthia LeBron, and Yumei Fu

Abstract

Murine embryonic stem cells (ES) are pluripotent cells and have the potential to become a wide variety of specialized cell types. Mouse ES cell differentiation can be regarded as a valuable biological tool that has led to major advances in our understanding of cell and developmental biology. In vitro differentiation of mouse ES cells can be directed to a specific lineage formation, such as hematopoietic lineage, by appropriate cytokine and/or growth factor stimulation. To study specific gene function in early developmental events, gene knockout approaches have been traditionally used; however, this is a time-consuming and expensive approach. Recently, we have shown that siRNA is an effective strategy to knockdown target gene expression, such as *Ape1*, during ES cell differentiation, and consequently, one can alter cell fates in ES-derived differentiated cells. This approach will be applicable to test the function of a wide variety of gene products using the ES cell differentiation system.

Key words: ES cell, RNAi, siRNA, *APE1*, stem cells, differentiation.

Abbreviations

EB	Embryoid body
siRNA	Small interfering RNA
HPC	Hematopoietic progenitor cells
dsRNA	Double-strand RNA
mIL-3	Murine interleukin-3
mSCF	Murine stem cell factor
mGM-CSF	Murine granulocyte-monocyte-colony stimulating factor
mFL	Murine Flt3 ligand
mLIF	Murine leukemia inhibitory factor
MACS	Magnetic associated cell sorting

B. Zhang, E.J. Stellwag (eds.), *RNAi and MicroRNA-Mediated Gene Regulation in Stem Cells*, Methods in Molecular Biology 650,
DOI 10.1007/978-1-60761-769-3_10, © Springer Science+Business Media, LLC 2010

1. Introduction

Murine ES (mES) differentiation is a robust system that can be used to study the regulation of hematopoietic cell development (1, 2). As observed in developing murine embryos in vivo, differentiated mES cells express similar cell surface antigens and molecular expression patterns at the appropriate stages of progenitor cell development. Mature blood cells, such as red blood cells, platelets, neutrophils, eosinophils, and natural killer cells, have been generated from mES cells (3–6). ES cells are able to form embryoid bodies (EBs) in the absence of LIF in culture (7). After dissociation of EBs to single cells by collagenase digestion, the EB cells can differentiate into myeloid progenitor cells when induced with appropriate cytokines (8, 9).

Ape1 is a multifunctional protein involved in DNA base excision repair activity, and in modulating DNA binding, including NF-kB, Egr-1, p53, HIF-1α, and Pax family, through redox mechanism (10). We recently demonstrated its role in mouse embryonic hematopoietic regulation (11) and regulation of pancreatic cancer cell growth and migration (12).

RNA interference (RNAi), a term coined by Fire and his colleagues, describes the inhibition of gene expression by double-stranded RNAs (dsRNAs) that have been introduced into worms (13). Guo and Kemphues (1995) first found that double-stranded RNA was more effective at producing interference of gene expression than either strand individually. After injection into adult *Caenorhabditis elegans*, single-stranded antisense RNA had a modest effect in diminishing specific gene expression, whereas double-stranded mixtures caused potent and specific interference (14). RNAi is a multistep process involving the generation of small interfering RNAs (siRNAs) in vivo through the action of the RNase III endonuclease Dicer. The resulting 21- to 23-nt siRNAs mediate degradation of their complementary RNA (15).

The traditional gene knockout techniques play a principal role in analyzing gene function during normal murine development; however, it is an expensive and time-consuming technique. Recently, siRNA has been used successfully to knockdown target gene expression in mammalian cells. ES differentiation is an attractive model for studying the molecular regulation of cell lineage commitment and cellular differentiation because ES cells give rise to cells derived from all three primary germ layers. Therefore, the ability to selectively knockdown specific target genes using siRNA would aid in the understanding of multiple aspects of early murine development. Our approach to knockdown Redox signaling protein *Ape1* gene expression in ES cell differentiation into hematopoietic cells is described.

2. Materials

2.1. Cells, Medium, and Serum

1. The CCE ES cells can be purchased from American Tissue Culture Collection (ATCC). CCE ES cells are derived from day 3 blastocysts of 129/SVJ mice, and maintain in culture with murine embryonic fibroblast feeder cells.
2. Iscove's Modified Dulbecco's Medium (IMDM) (12440-079, Invitrogen, Carlsbad, CA, USA).
3. Dulbecco's Modified Eagle's medium (DMEM) (10569, Invitrogen, Carlsbad, CA, USA).
4. Fetal bovine serum (FBS) (HCC6900, StemCell Technologies, Vancouver, Canada).

2.2. siRNA

1. Design Ape1 siRNA sequence: The Ape1 siRNA targeting sequence is GTCTGGTAAGACTGGAATACC (*see* **Notes 1** and **2**).
2. Synthesis of Ape1 siRNA: (*see* **Note 3**). Ape1 siRNA is synthesized commercially from Dharmacon Inc.
3. Control siRNA: Scrambled Ape1 siRNA (or shortly named scramble siRNA) can be purchased from Dharmacon Inc.

2.3. Antibodies

1. Rabbit antimouse Ape1 antibody (NB-100-909) can be purchased from Novus Biologicals (Littleton, CO) and goat antihuman-actin antibodies (SC-1615) purchased from Santa Cruz Biotechnology Inc (Santa Cruz, CA).

2.4. Cytokines

1. Murine leukemia inhibitory factor (mLIF) (02740, StemCell Technologies, Vancouver, Canada).
2. Murine interleukin-3 (mIL-3) (02733, StemCell Technologies, Vancouver, Canada).
3. Murine granulocyte-macrophage-colony stimulating factor (mGM-CSF), (02732, StemCell Technologies, Vancouver, Canada).
4. Murine stem cell factor (mSCF) (02731, StemCell Technologies, Vancouver, Canada).

2.5. Other Reagents

1. Anti-Biotin beads (130-091-147) are purchased from Miltenyi Biotec (Auburn, CA).
2. Oligofectamine 2000 (12252011) is purchased from Invitrogen (Carlsbad, CA).
3. Methylcellulose-based ES cell differentiation medium (M312D) and collagenase (07902) are purchased from StemCell Technologies (Vancouver, Canada).

4. Gelatin is purchased from StemCell Technologies (Vancouver, Canada).
5. Mouse embryonic fibroblasts (00321) are purchased from StemCell Technologies (Vancouver, Canada).
6. Collagnase (07902) is purchased from StemCell Technologies (Vancouver, Canada).
7. MACS buffer: PBS 500 mL supplemented with BSA 0.5 g, EDTA 2 mM. (Pass the solution to 0.22 μm filter before use).

3. Methods

3.1. In Vitro Maintenance of ES Cells

1. Mouse CCE ES cells are maintained on murine embryonic fibroblast feeder cells or gelatinized tissue culture dishes (100 mm; Costar, Cambridge, MA) in standard ES culture medium consisting of Dulbecco's modified Eagle's medium (DMEM) supplemented with 15% fetal calf serum (FCS; GIBCO, Grand Island, NY), 0.1 mmol L-glutamine, 150 mmol monothioglycerol (MTG), 100 U/mL penicillin, 100 mg/mL streptomycin, and 1,000 U/mL LIF (StemCell Inc., Vancouver, Canada).
2. The culture medium should be changed every day and the cells passaged every 2 or 3 days (16).

3.2. In Vitro Differentiation of ES Cells

1. Add CCE ES cells to 0.9% methylcellulose medium (StemCell Technologies, Vancouver, Canada), 15% FBS (fetal bovine serum, StemCell Technologies, Vancouver, Canada), 100 ng/mL stem cell factor (R&D System, Minneapolis, MN), and 450 μM monothioglycerol (Sigma, St. Louis, MO) at a cell concentration of 5,000~10,000 cells/mL plated in a 33-mm Petri dish. Efficient differentiation of ES cells to EBs occurs after 10 days of culture.
2. Harvesting EBs: Remove the EBs from methylcellulose by dilution with Iscove's Modified Dulbecco's Medium (IMDM).
3. Dissociate EB with collagenase: Add 3 mL collagenase to the EBs in the Facon tube, incubate at 37°C for 1 h.
4. Wash cells with IMDM medium.
5. Collect washed EB cell populations.

3.3. Preparation of dsRNA and Transfection of siRNA to EB Cells

1. Dilute EB cells with fresh medium without antibiotics and transfer to 6-well plates with 5×10^5 cells/well (500 μl per well). We perform a single transfection of siRNA duplex (Apel siRNA) using Oligofectamine 2000 Reagent and assay for gene silencing 2 days after transfection (*see* **Note 4**).

2. In a sterile 1.5-mL Eppendorf tube (tube A), add 50 nM Ape1 siRNA or control siRNA in Opti-MEM medium to reach a final volume of 50 μl. In another sterile 1.5-mL Eppendorf tube (tube B), add 3 μl of Oligofectamine 2000 to 12 μl of Opti-MEM medium (total 15 μl). Incubate tubes A and B separately for 10 min at room temperature.
3. Transfer the contents of tube B to tube A and mix by inversion 5 –10 times. Do not vortex the tube. Incubate the tube A (containing the mixture) for 25 min at room temperature. The solution will become turbid because siRNA binds to Oligofectamine 2000 to form a complex suspension.
4. Add 35 μl of fresh Opti-MEM medium to tube A to obtain a complex with a final volume of 100 μl.
5. Add the above-formed complex (100 μl) to the wells containing cells. Cells are maintained in a low serum conditions (2% serum) according to manufacturer's instruction at 50% confluence. Incubate about 4 h at 37°C.
6. Add an equal volume of culture medium with 18% serum to quench the transfection. Allow the cells to continue to grow for 3 days in a 5% CO_2 incubator.

3.4. Culturing Conditions for siRNA-Transfected EB Cells

1. EB cells were cultured with SCF, or with SCF, mGM-CSF, and mIL-3 (10 ng/mL) to promote myeloid differentiation.

3.5. Western Blot Analysis for Ape1

1. Isolate proteins from cultured cells after 72 h of siRNA treatment (*see* **Note 5**).
2. Ape1 Western blot analysis is performed using 20 μg of protein lysate. Proteins are separated by SDS-polyacrylamide electrophoresis using a 10% (w/v) polyacrylamide resolving gel and transferred electrophoretically to a nitro cellulose membrane.
3. Block membranes with 5% TBS/T (TBS containing 0.2% Tween 20) buffer for 1 h. Discard blocking solution and dilute Ape1 primary antibody (Novus Biologicals, LLC Littleton, CO) at a 1:400 dilution or the Beta-actin antibody at a 1:1000 dilution in 5% TBS/T (TBS containing 0.2% Tween 20). Immunoblotting with the primary antibody should be performed overnight at 4°C.
4. Primary antibody is removed and membrane washed three times for 5 min each with 5% TBS/T (TBS containing 0.2% Tween 20).
5. The peroxidase-conjugated secondary antibodies (Amersham Pharmacia Biotech) is diluted 1:1000 in blocking solution for 1 h at room temperature on a rocking platform.

6. All immunoblots were visualized by the Amersham electrochemi-luminescence (ECL) Advance Western Blotting Detection Kit according to the manufacturer's instructions (Amersham Pharmacia Biotech).

3.6. Immunocytochemistry for Ape1

1. Day 10 EB cells are cultured at 2×10^4 cells per well in a lab-Tek chamber slide system (Fisher Inc, Pittsburg, PA), in the presence of 10 ng/mL SCF.
2. Cells are transfected with Ape1 siRNA or Scrambled siRNA at 50 nM final concentration.
3. Cells are incubated for 72 h (*see* **Notes 6** and 7).
4. The cells are fixed with 4% paraformaldehyde for 5–10 min.
5. Ape1 antibody is added (1:400) to the cell slide and incubated overnight at 4°C with shaking.
6. The slide is washed with phosphate-buffered saline (PBS) for 10 min.
7. The Ape1 antimouse antibody is added and slide is incubated for 30 min (1:250) (Sigma, St. Louis, MO).
8. The slide is washed with PBS for 10 min.
9. The slide is exposed to Extravdin (1:250 dilution) for 3 h at room temperature.
10. The reaction product can be visualized with 0.05 DAB0.1 M phosphatase buffer0.01% H_2O_2 (Sigma, St. Louis, MO).
11. The positive cells can be scored under a microscope at 20X magnification microscope.

4. Notes

1. Dharmacon Research (Lafayette, CO) has been a valuable supplier of commercial RNAi synthesis in our experience.
2. Both Ape1 dsRNA and scrambled dsRNA were obtained from Dharmacon (Lafayette, CO). (To design the specifc siRNA sequence, go to siRNA designer software at the Dharmacon website: www.dharmacon.com). The key points to consider in selecting an siRNA sequence are as follows: (1) Start 75 bases downstream from the start codon; (2) locate the first AA dimer; (3) and record the next 19 nucleotides following the AA dimer; (4) subject the chosen 21-base sequence to a BLAST-search (NCBI) database to ensure that only one gene is targeted. A typical 0.2 μmol-scale RNA synthesis provides about 1 mg of RNA, which is sufficient for 1000 transfections.

3. We recommend designing siRNA with symmetric 3′TT overhang as previously recommended by Elbashir et al. (18) to facilitate an equal ratio of sense and antisense target RNA-cleaving siRNAs.
4. In siRNA transfection experiments, the efficiency of transfection may depend on the cell type, the passage number, and the confluency of the cells. Moreover, the time and the manner of formation of siRNAliposome complexes (e.g., inversion versus vortexing) are also critical. Please follow the instructions provided by the manufacturers. Low-transfection efficiencies are the most frequent cause of unsuccessful silencing. Good transfection is a nontrivial issue and needs to be carefully examined for each new cell line to be used.
5. Depending on the abundance and half-life (or turnover) of the target protein, a knockdown phenotype may become apparent after 1–3 days of siRNA transfection. If no phenotype is observed, depletion of the protein must be tested by immunofluorescence or Western blot analysis.
6. siRNAs introduced here usually are effective to knock-down the target gene expression and this effect is normally maintained for 3–5 days. After that period, siRNA will be degraded and the knockdown effect will be lost and target gene expression will recover in the cells.
7. For longer period of target gene-expression knockdown, generation and use of shRNA are discussed in detail by Hannon and Conklin (19).

Acknowledgments

The authors would like to thank Drs. Mervin C. Yoder and Mark Kelley at the Indiana University at Indianapolis for their collaboration in this work. This research was supported by CA 094025 and P30 CA82709.

References

1. Keller, G., Lacaud, G., and Robertson, S. (1999) Development of the hematopoietic system in the mouse. *Exp Hematol* **27,** 777–787.
2. Daley, G. Q. (2003) From embryos to embryoid bodies: Generating blood from embryonic stem cells. *Ann N Y Acad Sci* **996,** 122–131.
3. Hamaguchi-Tsuru, E., Nobumoto, A., Hirose, N., Kataoka, S., Fujikawa-Adachi, K., Furuya, M., and Tominaga, A. (2004) Development and functional analysis of eosinophils from murine embryonic stem cells. *Br J Haematol* **124,** 819–827.
4. Lieber, J. G., Webb, S., Suratt, B. T., Young, S. K., Johnson, G. L., Keller, G. M., and

Worthen, G. S. (2004) The in vitro production and characterization of neutrophils from embryonic stem cells. *Blood* **103,** 852–859.
5. Fujimoto, T. T., Kohata, S., Suzuki, H., Miyazaki, H., and Fujimura, K. (2003) Production of functional platelets by differentiated embryonic stem (ES) cells in vitro. *Blood* **102,** 4044–4051.
6. Nakayama, N., Fang, I., and Elliott, G. (1998) Natural killer and B-lymphoid potential in CD34+ cells derived from embryonic stem cells differentiated in the presence of vascular endothelial growth factor. *Blood* **91,** 2283–2295.
7. Potocnik, A. J., Kohler, H., and Eichmann, K. (1997) Hemato-lymphoid in vivo reconstitution potential of subpopulations derived from in vitro differentiated embryonic stem cells. *Proc Natl Acad Sci USA* **94,** 10295–10300.
8. Lu, S. J., Li, F., Vida, L., and Honig, G. R. (2002) Comparative gene expression in hematopoietic progenitor cells derived from embryonic stem cells. *Exp Hematol* **30,** 58–66.
9. Nakayama, N., Lee, J., and Chiu, L. (2000) Vascular endothelial growth factor synergistically enhances bone morphogenetic protein-4-dependent lymphohematopoietic cell generation from embryonic stem cells in vitro. *Blood* **95,** 2275–2283.
10. Merluzzi S, Moretti M, Altamura S, Zwollo P, Sigvardsson M, Vitale G, Pucillo C. (2004) CD40 stimulation induces Pax5/BSAP and EBF activation through a APE/Ref-1-dependent redox mechanism. *J Biol Chem* **279**(3), 1777–1786.
11. Zou GM, Luo MH, Reed A, Kelley MR, Yoder MC. (2007) Ape1 regulates hematopoietic differentiation of embryonic stem cells through its redox functional domain. *Blood* **109**(5), 1917–1922.
12. Zou, G.M., Maitra, A. (2008) Small-molecule inhibitor of the AP endonuclease 1/REF-1 E3330 inhibits pancreatic cancer cell growth and migration. *Mol Cancer Ther* 7(7), 2012–2021.
13. Fire, A., Xu, S., Montgomery, M. K., Kostas, S. A., Driver, S. E., and Mello, C. C. (1998) Potent and specific genetic interference by double-stranded RNA in *Caenorhabditis elegans. Nature* **391, 806–811.**
14. Guo, S., and Kemphues, K. J. (1995) par-1, a gene required for establishing polarity in *C. elegans* embryos, encodes a putative Ser/Thr kinase that is asymmetrically distributed. *Cell* **81,** 611–620.
15. Shi, Y. (2003) Mammalian RNAi for the masses. *Trends Genet* **19,** 9–12.
16. Zou, G. M., Reznikoff-Etievant, M. F., Hirsch, F., and Milliez, J. (2000) IFN-gamma induces apoptosis in mouse embryonic stem cells, a putative mechanism of its embryotoxicity. *Dev Growth Differ* **42,** 257–264.
17. Zou, G. M., Chen, J. J., Yoder, M. C., Wu, W., and Rowley, J. D. (2005) Knockdown of Pu.1 by small interfering RNA in CD34+ embryoid body cells derived from mouse ES cells turns cell fate determination to pro-B cells. *Proc Natl Acad Sci U S A* **102,** 13236–13241.
18. Elbashir, S. M., Harborth, J., Lendeckel, W., Yalcin, A., Weber, K., and Tuschl, T. (2001) Duplexes of 21-nucleotide RNAs mediate RNA interference in cultured mammalian cells. *Nature* **411,** 494–498.
19. Hannon, G. J., and Conklin, D. S. (2004) RNA interference by short hairpin RNAs expressed in vertebrate cells. *Methods Mol Biol* **257,** 255–266.

Chapter 11

Proteins Involved in Cell Migration from Glioblastoma Neurospheres Analyzed by Overexpression and siRNA-Mediated Knock-Down

Carsten Hagemann, Harun M. Said, Michael Flentje, Klaus Roosen, and Giles Hamilton Vince

Abstract

Glioblastoma multiforme (GBM) are the most common malignant brain tumours in adults, characterized by short survival periods of patients. Their aggressive local growth pattern and increased invasiveness, due to a high motility of the tumour cells, hamper treatment. However, the molecular mechanisms regulating glioblastoma cell migration are still elusive. Here, we describe the combination of a highly efficient cell transfection by nucleofection® technology and the generation of spheroids from these transfected glioblastoma cell lines. Nucleofection allows the manipulation of protein expression by overexpression and siRNA-mediated protein knock-down. Transfection efficiencies >80% can be achieved with some GBM cell lines. Transfected neurospheres then can be used for migration assays (as described here in detail) and a multitude of other functional assays. In comparison to monolayer cultures, the advantage of spheroids is their resemblance to organized tissue in combination with the accuracy of in vitro methodology and marked experimental flexibility.

Key words: Glioblastoma multiforme, glioma, migration assay, neurosphere, nucleofection, overexpression, siRNA, spheroid, transfection, Western blotting.

1. Introduction

1.1. Human Malignant Gliomas

Glioblastoma multiforme (GBM) are the most prevalent and highly malignant brain tumours of adults (1). They may develop from low-grade astrocytoma (WHO grade II) and anaplastic astrocytoma (WHO grade III), respectively, or appear de novo without any precursor lesion (1). The median survival of patients

B. Zhang, E.J. Stellwag (eds.), *RNAi and MicroRNA-Mediated Gene Regulation in Stem Cells*, Methods in Molecular Biology 650, DOI 10.1007/978-1-60761-769-3_11,

is very low and averages to only 14.6 months, despite multidisciplinary treatment including surgery, chemotherapy and concomitant γ-irradiation (2). This poor prognosis can be traced back to an aggressive local growth pattern and a high degree of diffuse infiltration of the adjacent brain parenchyma by tumour cells (3). Therefore, complete tumour resection – which would be highly advantageous for the patient (4) – is rarely achieved and recurrence of GBM occurs regularly (3). Although several factors involved in regulation of GBM cell motility have been identified (3), the molecular mechanisms, which would allow a targeted therapeutic interference, are largely unknown.

1.2. Use of Neurospheres to Study Glioma Cell Migration and Invasion

Different model systems are in use to study glioma cell migration and invasion, e.g., the simple scratch technique in a confluent cell monolayer, the use of cloning rings and the Boyden chamber technique (5). However, discrepancies have been observed when drug effects were studied in parallel on monolayer cultures in vitro and on solid tumours in vivo (6, 7). Seeding cells onto plates base-coated with Noble Agar promotes formation of three-dimensional spheroids, because there is no appropriate surface for cell attachment (8). Such tumour cell spheroids share characteristics of in vivo tumours, like cell-cell contacts involving extracellular matrix, chronically hypoxic populations of deeper lying cells, necrotic regions in the centre and cell cycle times that range from exponential proliferation rates of peripheral cells through essentially nondividing, resting cells deeper in the spheroid (8–10). Therefore, they combine the relevance of organized tissues with the accuracy of in vitro methodology (7, 10). Since individual spheroids can be used in long-term studies (up to 2 months) in medium volumes as small as 1 mL and as they are easy to handle, they offer a broad range of experimental flexibility (8). In neurooncological research, such approaches encompass tests of γ-irradiation efficiency (10), evaluation of chemotherapeutics (11), confrontation assays to study invasion into spheroids of reaggregated foetal brain (11) and even implantation into the brain of rats for in vivo treatment studies (12, 13).

1.3. Transfection of Cells by Nucleofection® Technology

The introduction of DNA and RNA into cells is an important technique in molecular biology to manipulate protein concentrations within the cell by transient overexpression and siRNA-mediated protein knock-down of specific signalling proteins. Ideally, these manipulations lead to phenotypic changes of cells, which will allow conclusions concerning the function of the protein.

The most broadly used transfection method is lipofection (14). However, cytotoxicity is a problem and transfection efficiencies of 40% or greater are seldom reached (15). Such transfection efficiencies are too low to use transfection in combination

Table 11.1
Transfection efficiencies of GBM cell lines using nucleofector technology, Solution V and Program T-20

Cell line	Transfection efficiency (%)	Viability (%)
U87 MG[a]	58	82
U251 MG	63	95
U138 MG	71	44
U343[a]	59	70
U373 MG	71	95
GaMG[a]	83	97
A1207	62	71
T5135	69	79
T6217	59	89
T3868[b]	55	77
TX3868[b]	53	72
Primary cells	81	98

[a]This cell line does not form proper spheroids after transfection using these conditions. But further optimisation may improve neurosphere formation.
[b]This cell line does not form spheroids, neither untransfected, nor transfected.
For source of cell lines used, *see* **Note 1**.

with spheroid based assays. Nucleofection® technology (Lonza Cologne, formerly Amaxa) is an electroporation-based technique that offers high transfection efficiencies in combination with low cell mortality (**Table 11.1**). It does not cause alterations of the cell's phenotype (15) and transfected GBM cells can be used to generate neurospheres (**Fig. 11.1**). However, high transfection efficiencies are required for experiments with spheroids composed of transfected cells. Nucleofection achieves its high transfection efficiencies by direct introduction of DNA into the nucleus of the cells without the necessity of cell division (16, 17). The DNA is bound to proteins containing a nuclear localisation signal and this complex is then transferred into the cell by electroporation (16). The DNA-protein complex is actively transported by the cell into its nucleus, which allows gene expression of exogenous DNA shortly after transfection. However, each cell line and primary cell culture has its own characteristics, including a unique composition of the cell membrane. Therefore, the parameters of the electric field for electroporation and the suitable transfection solution have to be determined for every cell type by an optimisation process (15). Recently, we optimised transfection of the GBM cell line U251 MG and measured transfection efficiencies of other GBM cell lines using the same conditions (**Table 11.1**)

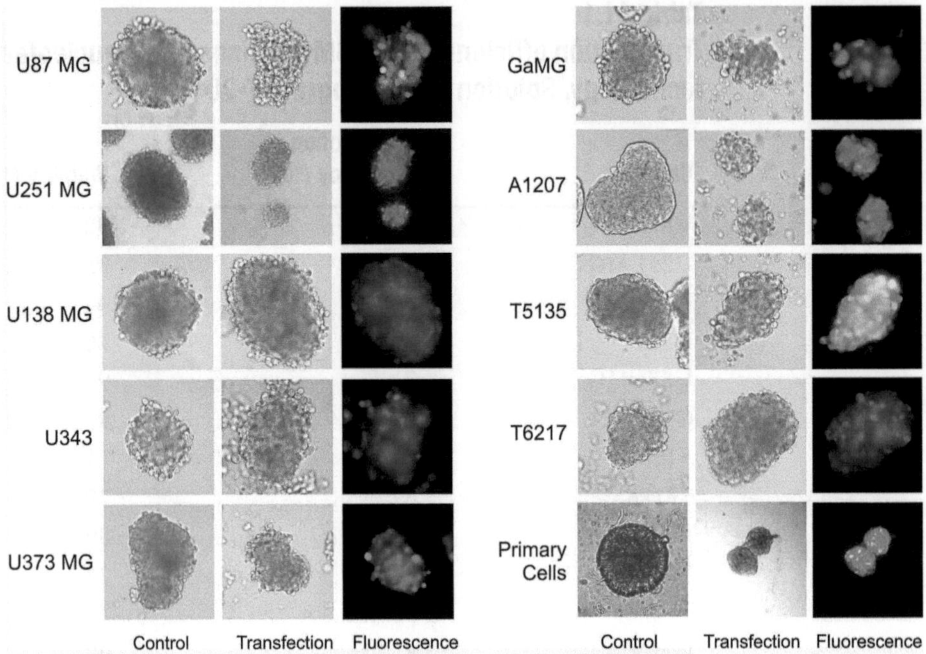

Fig. 11.1. Spheroids formed by untransfected (control) and transfected GBM cell lines. The different cell lines were nucleofected as described using 2.5 μg pmaxGFP (Lonza Cologne), a plasmid encoding green fluorescent protein. Forty-eight (48) h after transfection, expression of GFP was visualized using a fluorescence microscope (fluorescence). U87 MG, U343 and GaMG cells do not form proper spheroids after transfection using conditions which had been optimised for U251 MG cells (15). To run further optimisation may improve neurosphere formation by these cell lines.

(15). Nevertheless, running a full optimisation routine may detect even better conditions for these cells.

Here, we describe transfection of glioblastoma cell lines by nucleofection with plasmids for transient protein overexpression and siRNA-mediated protein knock-down, respectively. In addition, we describe generation of neurospheres from these transfected cells, the directional migration of cells from the spheroid on differently coated surfaces (11) and finally, detection of successful protein overexpression or knock-down by Western blotting.

2. Materials

2.1. Cell Culture and Nucleofection

1. For U87 MG, U251 MG, U138 MG, U373 MG, U343 and GaMG: Dulbecco's Modified Eagle's Medium (DMEM) containing 1 g/L glucose, sodium pyruvate and 3.7 g/L $NaHCO_3$, without L-glutamine supplemented with 10% heat-inactivated foetal calf serum (FCS), 2× nonessential

amino acids (NEA, 100× stock, add 10 mL NAE to 500 mL medium), 3 mM L-glutamine, 100 U/mL penicillin and 100 mg/mL streptomycin.

2. For A1207: RPMI-1640 containing L-glutamine supplemented with 10% FCS, 100 U/mL penicillin and 0.1 mg/mL streptomycin.
3. For T3868: DMEM containing 4.5 g/L glucose, without pyruvate, with L-glutamine supplemented with 10% FCS, 100 U/mL penicillin and 0.1 mg/mL streptomycin. For culture of TX3868, add 2 mM Glutamax.
4. Vent cap canted neck cell culture flasks (75 cm^2).
5. Phosphate-buffered saline (PBS) without Ca^{2+} and Mg^{2+}, low endotoxin.
6. 0.25% Trypsin-EDTA.
7. Nucleofector® device and Nucleofector Solution Kit V (Lonza Cologne, formerly Amaxa, Cologne, Germany) (*see* **Note 2**).

2.2. Generation of Neurospheres

1. Cell culture medium as stated above (**Section 2.1**), appropriate to the cell line used.
2. Ultra pure water.
3. Noble Agar.
4. 6-well cell culture cluster, flat bottom with lid.

2.3. Migration Assay

1. 96-well cell cluster, flat bottom with low evaporation lid.
2. Poly-L-lysine 0.01% solution.
3. Cell culture medium as stated above (**Section 2.1**), appropriate to the cell line used.
4. Cell culture plates (6 cm).

2.4. Cell Lysis, SDS-Polyacrylamide Gel Electrophoresis and Western Blotting

1. PBS without Ca^{2+} and Mg^{2+}, low endotoxin.
2. Protein lysis buffer: 10 mM Tris-HCl, pH 7.4, 150 mM NaCl, 1 mM ethylenediaminetetraacetic acid (EDTA), 1 mM ethylene glycerol-bis (β-aminoethylether)-N,N,N′,N′-tetraacetic acid (EGTA), 1% (v/v) Triton X-100 and 0.5% (v/v) IGEPAL CA-630. Store the buffer at 4°C. Before use add protease-inhibitors: 1 mM phenylmethanesulfonylfluoride (PMSF) (*see* **Note 3**), 10 μg/mL leupeptin (*see* **Note 4**), 23 μg/mL aprotinin. Handle these substances with care, since they are toxic.
3. Invitrogen Western blotting system consisting of: X Cell SureLock™ electrophoresis midi system and blot module, NuPAGE® SDS buffer kits (MES/MOPS) for

Bis-Tris polyacrylamide gels, Tris-acetate buffer for Tris-acetate gels, precasted polyacrylamide gels, antioxidant and transfer buffer (*see* **Note 5**).

4. Prestained molecular weight markers SeeBlue Plus 2 for NuPAGE Bis-Tris gels cover a range from 3 to 188 kDa in combination with MES buffer and from 14 to 191 kDa in combination with MOPS buffer. HiMark™ is used for Tris-acetate gels and covers a range from 31 to 460 kDa. These protein standards are from Invitrogen. However, numerous competitive reagents are available from other commercial sources.
5. Methanol.
6. Distilled water.
7. Protran® nitrocellulose transfer membrane and Whatman® paper filter sheets.
8. 1× Tris-buffered saline with Tween (TBST): 50 mM Tris base, 150 mM NaCl, 0.1% (v/v) Tween-20 (*see* **Note 6**).
9. Blocking buffer: 5% (w/v) nonfat dry milk in TBST.
10. Primary and secondary antibodies according to the protein of interest.
11. Enhanced chemiluminescent (ECL) reagent and Amersham Hyperfilm™ ECL.
12. Two acetate sheet protectors and an X-ray film cassette.
13. Stripping buffer: 60 mM Tris-HCl, pH 6.8, 200 mM β-Mercaptoethanol, 2% (w/v) SDS.

3. Methods

3.1. Cell Culture and Transfection of Cells by Nucleofection

1. GBM cell lines are maintained in 75-cm^2 cell culture flasks, containing 15–20 mL of the appropriate cell culture medium, at 37°C, 5.0% CO_2 and 100% humidity. Cells are passaged when they reach subconfluency with trypsin-EDTA.
2. Two days prior to transfection, passage the cells (*see* **Note 7**) and then grow them to 80% confluency.
3. For nucleofection, remove the culture medium with a pipette, wash the cell monolayer once with 10 mL PBS and add 3 mL of trypsin-EDTA solution. Incubate for 5 min at 37°C. Check microscopically to determine whether all cells are detached and add 7 mL of the appropriate culture medium.

4. Count the cells. For each nucleofection reaction 1×10^6 cells are required (*see* **Note 8**).
5. Fill the required amount of cell suspension into a 15-mL tube and centrifuge at 1,100×*g* for 10 min. Turn off the break. Wash the sedimented cells once with PBS.
6. For each 1×10^6 cells, add 100 μL of Nucleofection solution V (*see* **Note 8**), prewarmed to room temperature and resuspend cells gently. If several transfection reactions are planned, aliquot cells into Eppendorf-cups so that the cups contain 100 μL of solution V with 1×10^6 cells, each (*see* **Note 9**).
7. It is important to complete this whole step for each sample separately. Add 2 μg of plasmid DNA (in a maximal volume of 5 μL) or 3 μg of siRNA and mix gently (*see* **Note 10**). Transfer the cells to the electroporation cuvette making sure that the sample covers the bottom of the cuvette and avoid air bubbles while pipetting. Close the cuvette with the blue cap and place it into the Nucleofector device. Perform nucleofection using program T20 (*see* **Note 8**). Immediately after electroporation, add 500 μL of prewarmed appropriate cell culture medium to the cuvette.

3.2. Generation of Neurospheres from Transfected Cells

1. For preparation of agar-coated 6-well cell culture clusters, mix 20 mL ultra-pure water with 1 g Noble Agar and boil the mixture in a microwave oven to melt the agar. Add 80 mL of prewarmed (37°C) appropriate cell culture medium to the melted agar and mix using a pipette. Immediately, transfer 2 mL of the mixture to each well.
2. Wait until the agar has hardened at room temperature and store the clusters at 4°C for up to 2 weeks.
3. When performing nucleofection, prewarm the 6-well cell culture cluster at 37°C and add 4.5 mL of prewarmed (37°C) appropriate cell culture medium per well.
4. Transfer the transfected cells from the cuvettes to the wells (*see* **Note 11**). Work rapidly and cautiously, since cells are very sensitive after transfection. Incubate the cells at 37°C, 5% CO_2 and 100% humidity for 24 –48 h. Spheroids should form during this time.

3.3. Migration Assay

1. Add 50 μL of poly-L-lysine per well into a 96-well cell cluster. Incubate at 37°C for 30 min. Remove the Poly L-lysine from the wells and dry the plate for 30 min at 60°C (*see* **Note 12**).
2. Add 200 μL of appropriate cell culture medium per well and incubate at 37°C.

3. Evaluate the size of spheroids by means of an ocular scale operated by a micrometer screw on an inverted microscope. The spheroids should have reached a size of 200–300 μm. Use only well-formed spheroids without necrosis (**Fig. 11.1**). Carefully transfer 2 mL of the spheroid culture from the agar-coated well to an uncoated 6-cm cell culture plate. Under visual or microscopic control pick spheroids from the cell culture plate using a glass Pasteur pipette, connected to a Glasfirn PiPump, and place a single spheroid into the middle of each well of the 96-well cell cluster (*see* **Note 13**). Do not discard spheroids not needed for the assay, since these are required for Western blot analysis.
4. Take photos of the spheroid and migrating cells at time points 0 , 12 , 24 , 36 and 48 h (**Fig. 11.2**). Always use the same magnification. A simple way to evaluate migration is by measuring the area covered by cells spreading out from the spheroid. Determine the orthogonal diameters of the colonies at selected time-points and compare the values of different experimental settings (**Fig. 11.2**).

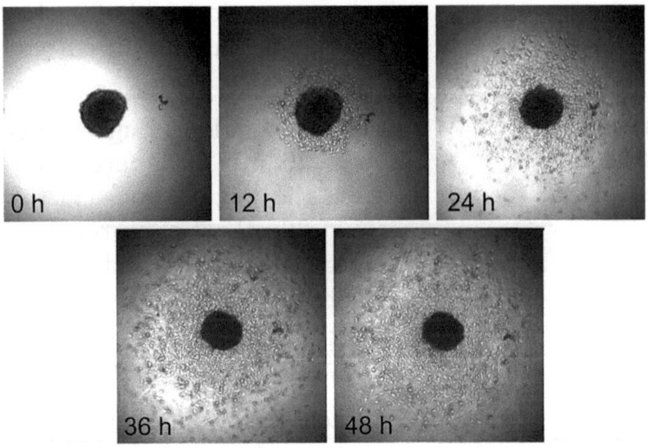

Fig. 11.2. Example of a migration assay. A spheroid, formed by transfected U251 MG cells, was placed into the middle of a Poly-L-lysine-coated well of a 96-well cell cluster. Photographs were taken 0, 12, 24, 36 and 48 h after positioning the neurosphere as indicated. Cells spreading out from the spheroid are clearly visible, enlarging the area covered by cells with progressing time due to migration.

3.4. Western Blot Detection of Protein Overexpression and Knock-Down, Respectively

1. Spheroids not needed for the migration assay are transferred into a 15-mL tube. The spheroids will settle to the bottom of the tube by gravity, do not centrifuge. Carefully remove the supernatant and wash spheroids twice with PBS. Do not centrifuge, let spheroids settle by gravity.
2. Remove the PBS and add 50 μL of ice-cold protein lysis buffer containing protease inhibitors. Mix thoroughly and then transfer the cell suspension into a 1.5-mL tube. Work on ice. Centrifuge for 10 min with 20,000×*g*. Transfer the

supernatant (protein lysate) to a new tube and discard the pellet.

3. These instructions are based on the use of the Invitrogen Western blotting system (*see* **Note 5**). Mix 32.5 μL of protein lysate with 12.5 μL protein loading buffer and 5 μL sample reducing agent. Heat for 10 min at 70°C. Use 25 μL per lane of precasted gels.
4. Prepare the gels (choose Bis-Tris, Tris-acetate and percentage of gel according to the size of the protein of interest, *see* **Note 5**) by removing the plastic lid from the top and the paper strip from the bottom edge. Remove the sample well-former carefully without damaging the gel-pockets and fix the gel-cassette in the X Cell SureLock^{TM} electrophoresis midi system with pockets showing to the inside. If electrophoresing only one gel, use a plastic place-holder to substitute for the second gel. Lock the tension wedge.
5. Rinse the gel-pockets with running buffer and load the gels with 25 μL sample per lane (*see* **Note 14**).
6. Fill the inner buffer-chamber with 200 mL of 1× MOPS, MES, or Tris-acetate electrophoresis buffer, according the protein size to be separated and the gel used (*see* **Note 5**). Add 500 μL of antioxidant to this buffer. Fill the outer buffer chamber with 600 mL electrophoresis buffer without antioxidant.
7. Connect to a power supply and electrophorese for 1 h at 200 V when using MOPS or MES buffer and 1 h at 150 V when using Tris-acetate buffer, respectively.
8. Prepare 1× NuPAGE transfer buffer: For one gel, mix 50 mL of 20× NuPAGE transfer buffer with 100 mL methanol, 1 mL NuPAGE antioxidant and 750 mL distilled water; for two gels, use 200 mL methanol and only 650 mL distilled water.
9. Remove the gels from the electrophoresis chamber, place the gel knife in the gap at the side edge of the gel-cassette and remove the plastic cover by pushing up and down the gel knife handle. Remove the loading pockets and wash the gel in transfer buffer.
10. Prepare the nitrocellulose transfer membrane and Whatman filter sheets in the size of the provided blotting pads. Soak the membrane, the filter sheets and the blotting pads in transfer buffer. Assemble the "blotting sandwich" submerged in a tray filled with transfer buffer: Filter paper, gel, transfer membrane, filter paper. Place two blotting pads into the blot module and put the sandwich on top, such that the nitrocellulose membrane is between the gel and the anode. It is vitally important to ensure this orientation or the protein will be lost from the gel into the buffer rather

than transferred to the nitrocellulose. Remove air bubbles from the sandwich by rolling a Pasteur pipette over the surface with weak pressure and place two blotting pads on top of the sandwich. For blotting of two gels, the final sandwich looks like this: Blotting pad, blotting pad, filter paper, first gel, transfer membrane, filter paper, blotting pad, filter paper, second gel, transfer membrane, filter paper, blotting pad, blotting pad.

11. Place the blot module into the chamber, fix it with the gel tension wedge and fill the blot module with transfer buffer. Fill the chamber with 600 mL of distilled water, connect to a power supply and electrophorese at 30 V for 1 h if blotting one gel and 2 h if blotting two gels, respectively.
12. Once the transfer is complete, take the blot module out of the chamber and disassemble it carefully. The prestained molecular weight markers should be clearly visible on the membrane. Since the staining may wash out during the following procedures, mark marker bands with a ball pen.
13. Incubate the nitrocellulose membrane in blocking buffer on a rocking platform for 30 min.
14. Add primary antibody to 2 mL of blocking buffer. The amount of antibody necessary depends on the antibody and batch used and usually is given in the manufacturer's manual. Otherwise, try different dilutions in preceding optimisation experiments.
15. Transfer the nitrocellulose membrane to a 50-mL tube and add the 2 mL antibody solution. Close the tube with its lid and incubate it on a roller at 4°C for at least 1.5 h or overnight. The incubation time depends on the antibody used.
16. Wash the nitrocellulose 3× for 5 min each with TBST in a tray placed onto a rocking platform.
17. Add secondary antibody to 2 mL TBST. The dilution of the secondary antibody varies depending on different manufacturers and batches. The recommended dilutions can be found in the data-sheets provided. Transfer the solution into a fresh 50-mL tube. Place the nitrocellulose into the tube, close the lid and incubate it on a roller for 1 h at 4°C.
18. Discard the secondary antibody and wash the nitrocellulose 5× for 5 min each with TBST in a tray on a rocking platform.
19. Dry the blot carefully by dabbing with a hygienic tissue. Place the blot between the leaves of an acetate sheet protector. Mix 1 mL aliquots of each portion of the ECL reagent and pipette the mixture onto the blot. Close the cover

leaf of the acetate sheet protector, remove air bubbles by squeezing them out by hand and incubate for 2 min.

20. Place the blot between the leaves of a fresh acetate sheet protector that has been cut to the size of an X-ray film cassette and place it in such a cassette. In a dark room under safety light conditions, position a Hyperfilm on top of the acetate and incubate for a suitable exposure time, typically between 30 s and 5 min.
21. Once a satisfactory exposure for the protein of interest has been obtained, the membrane can be stripped of the probe and reprobed with an antibody that recognizes a suitable housekeeping protein as loading control, e.g. γ-Tubulin, β-Actin or GAPDH.
22. For stripping the blots, incubate the membrane for 1–2 h in stripping buffer on a rocking platform. Perform this step in a closed plastic box in a fume cabinet. After stripping the blot, wash it 7× for 5 min each with TBST. Then continue with Step 13 above. Blots can be stripped several times; however, signal strength will weaken each round.

4. Notes

1. Several different GBM cell lines were established from patients. U87 MG originated from a 44-year-old Caucasian female (18, 19). U251 MG were derived from a 75-year-old Caucasian male (20, 21), U138 MG from a 47-year-old Caucasian male (18, 19) and U373 MG from a 61-year-old Caucasian male (19). These cell lines grow in an epithelial, fibroblast-like morphology and were originally purchased from ATCC (American Type Culture Collection, Rockville, MD), as were U343. GaMG were established from the GBM of a 42-year-old Caucasian female by the Gade Institute of the University Bergen, Norway (22). They are adherent, large, spindlelike cells. A1207 (originally from ATCC), T5135, T6217, T3868 and TX3868 were a kind gift by Samuel Samnick (Department of Nuclear Medicine, Experimental Nuclear Medicine, University Hospital Würzburg, Germany). TX3868 was established from a xenograft of human T3868 GBM cells into nude mice (23). These two cell lines, T5135 and T6217, were established in the Departments of Human Genetics and Neurosurgery, Saarland University, Germany.
2. The Nucleofector kit contains the transfection solution and supplement. Before performing the first transfection, add

the supplement to the transfection solution (as stated in the manufacturer's manual) and mix gently. When both are added, the solution can be stored at 4°C for up to 3 months. Note the date on the vial. The kit also includes nucleofection cuvettes and plastic pipettes. Do not reuse either of these, as repeated use may diminish transfection efficiency.

3. Dissolve 10 mg/mL PMSF in methanol and store in aliquots at 4°C. The solution is stable for up to 9 months. Use 100 μL for 10 mL of protein lysis buffer.
4. Dissolve 0.5 mg/mL Leupeptin in water and store in aliquots at –20°C. The solution is stable for up to 6 months. Use 10 μL for 10 mL of protein lysis buffer.
5. Other Western blotting systems may be as suitable as the system described here. However, Invitrogen offers the fine-tuned NuPAGE Western blotting system, including pre-cast gels and buffer kits. Our laboratory found this system very useful, because it is easy and safe to handle and produces reproducible results. Bis-Tris gels are available in three acrylamide concentrations: 10%, 4–12% and 12%. They provide good separation and resolution of small- to medium-sized proteins up to 200 kDa when using a neutral pH buffer system. The system is based on Bis-Tris-HCl buffered polyacrylamide gels (pH 6.4), with separation gels that operate at pH 7.0. Although they do not contain SDS, they are formulated for denaturing gel electrophoresis only. By combining any of these three gel types with the NuPAGE MES or MOPS buffers, six separation ranges can be obtained. The MES buffer is recommended for resolving small proteins, while MOPS buffer is useful for resolving medium- to large-size proteins. For large proteins up to 400 kDa, Tris-acetate gels should be used in combination with Tris-acetate buffer. Invitrogen offers a chart that helps in selection of the optimal conditions for protein separation. All gels can be stored for up to 1 year at temperatures between 4 and 25°C. Use of NuPAGE LDS sample buffer and NuPAGE sample reducing agent will ensure complete sample reduction. For transfer, NuPAGE transfer buffer is recommended, since all components of this system are matched. All buffers are stored at 4°C.
6. Usually, we prepare a stock of 10 L 1× TBST and store it at room temperature.
7. A cell density >80% may cause lower nucleofection efficiencies. To reach 80% confluency on the day of transfection, split U87 MG, U251 MG, U373 MG and GaMG in a ratio of 1:10, U138 MG, U343 and A1207 in a ratio of 1:3

and T3868 and TX3868 in a ratio of 1:2. Since slightly different culture conditions may influence proliferation of cells, it is advisable to test cell growth rates before planning nucleofection. Make sure that agar-coated plates for generation of spheroids are prepared in time, as described in **Section 3.2**.

8. The optimum cell number, Nucleofector solution, amount of DNA/siRNA and Nucleofector program are cell-type specific and have to be determined in an optimisation process (15). For details see the website of Lonza Cologne (http://www.lonzabio.com). Generally, 1×10^6 cells are used per transfection. However, cell numbers between 5×10^5 and 5×10^6 may give high-transfection efficiencies using the method described in this chapter. For lower cell numbers between 2×10^5 and 4×10^5, a special 96-well Nucleofector device is available, which requires modified protocols and kits. Three different Nucleofector solutions are offered. For GBM cell lines, we found solution V to give the best results without influencing the cell's phenotype in combination with program T-20 (15).
9. Do not try to transfect too many samples in one pass, because cells should not be stored for more than 15 min in Nucleofector solution, as this reduces cell viability and gene-transfer efficiency.
10. The quality and the concentration of DNA used for nucleofection play a central role in the efficiency of gene transfer. Therefore, plasmid DNA should be highly purified and dissolved in deionized water. Different plasmid DNAs can be mixed to perform double transfection. However, a maximum of 5 μg in a total volume of 5 μL should not be exceeded. In case the transfection efficiency is too low, the amount of DNA can be optimised by titration. The amount of siRNA for optimum protein knockdown may vary depending on the siRNA sequence and the target protein and has to be evaluated in preliminary experiments.
11. Alternatively, cells can be plated on uncoated 6-well cell culture clusters to grow them as transfected monolayers.
12. The migration behaviour of cells may vary, dependent on the surface used for the migration assay. Therefore, instead of Poly-L-lysine, other matrix proteins like laminin, fibronectin or collagen can be used for base-coating. Dissolve these extracellular matrix proteins in a concentration of 20 μg/mL in PBS and coat the 96-well cell cluster as described. Alternatively, migration on uncoated plates also can be measured.

13. Using Pasteur glass pipettes, a visual control of the collected spheroids inside the pipette is possible. In our hands, the Glasfirn PiPump allows a more precise manipulation compared to other pipettors. However, picking and transferring spheroids require some amount of training. Therefore, it is advisable to practice this procedure before starting an experiment.
14. It is important to load the gel before filling the chambers with buffer. Load the gel asymmetrically, e.g. load the marker in the left lane followed by the samples to the right. By using this loading strategy, the prestained and visible marker will reveal the orientation of the gel. Lanes not needed for samples should be filled with a few microlitres of loading buffer to avoid "smiling".

Acknowledgements

We are very grateful to Stefanie Gerngras and Siglinde Kühnel for technical assistance. This project was supported by Interdisziplinäres Zentrum für Klinische Forschung der Universität Würzburg (CH and GHV).

References

1. Reifenberger, G., and Collins, V. P. (2004) Pathology and molecular genetics of astrocytic gliomas. *J Mol Med* **82**, 656–670.
2. Stupp, R., Mason, W. P., van den Bent, M. J., Weller, M., Fisher, B., Taphoorn, M. J., Belanger, K., Brandes, A. A., Marosi, C., Bogdahn, U., Curschmann, J., Janzer, R. C., Ludwin, S. K., Gorlia, T., Allgeier, A., Lacombe, D., Cairncross, J. G., Eisenhauer, E., Mirimanoff, R. O., European Organisation for Research and Treatment of Cancer Brain Tumor and Radiotherapy Groups, and National Cancer Institute of Canada Clinical Trials Group (2005) Radiotherapy plus concomitant and adjuvant temozolomide for glioblastoma. *N Engl J Med* **352**, 987–996.
3. Demuth, T., and Berens, M. E. (2004) Molecular mechanisms of glioma cell migration and invasion. *J Neur-Oncol* **70**, 217–228.
4. Stummer, W., Reulen, H.-J., Meinel, T., Pichlmeier, U., Schumacher, W., Tonn, J.-C., Rohde, V., Oppel, F., Turowski, B., Woiciechowsky, C., Franz, K., Pietsch, T., and ALA-Glioma Study Group (2008) Extent of resection and survival in Glioblastoma multiforme: Identification of and adjustment for bias. *Neurosurgery* **62**, 564–576.
5. Pilkington, G. J., Bjerkvig, R., De Ridder, L., and Kaaijk, P. (1997) In vitro and in vivo models for the study of brain tumour invasion. *Anticancer Res* **17**, 4107–4109.
6. Martin, W. M. C., and McNally, N. J. (1980) Cytotoxicity of Adriamycin to tumour cells in vivo and in vitro. *Br J Cancer* **42**, 881–889.
7. Nederman, T. (1984) Effects of Vinblastine and 5-Fluorouracil on human glioma and thyroid cancer monolayers and spheroids. *Cancer Res* **44**, 254–258.
8. Yuhas, J. M., Li, A. P., Martinez, A. O., and Ladman, A. J. (1977) A simplified method for production and growth of multicellular tumor spheroids. *Cancer Res* **37**, 3639–3643.

9. Sutherland, R. M., McCredie, J. A., and Inch, W. R. (1971) Growth of multicell spheroids in tissue culture as a model of nodular carcinomas. *J Natl Cancer Inst* **46**, 113–120.
10. Sutherland, R. M. (1980) The multicellular spheroid system as a tumor model for studies of radiation sensitizers. *Pharmac Ther* **8**, 105–123.
11. Engebraaten, O., Bjerkvig, R., and Berens, M. E. (1991) Effect of alkyl-lysophospholipid on glioblastoma cell invasion into fetal rat brain tissue in vitro. *Cancer Res* **51**, 1713–1719.
12. Farrell, C. L., Stewart, P. A., and Del Maestro, R. F. (1987) A new glioma model in rat: The C6 spheroid implantation technique permeability and vascular characterization. *J Neurooncol* **4**, 403–415.
13. Goldbrunner, R. H., Wagner, S., Roosen, K., and Tonn, J. C. (2000) Models for assessment of angiogenesis in gliomas. *Neurooncol* **50**, 53–62.
14. da Cruz, M. T., Simoes, S., and de Lima, M. C. (2004) Improving lipoplex-mediated gene transfer into C6 glioma cells and primary neurons. *Exp Neurol* **187**, 65–75.
15. Hagemann, C., Meyer, C., Stojic, J., Eicker, S., Gerngras, S., Kühnel, S., Roosen, K., and Vince, G. H. (2006) High efficiency transfection of glioma cell lines and primary cells for overexpression and RNAi experiments. *J Neurosci Meth* **156**, 194–202.
16. Christine, R., and Siebenkotten, G. (2000) Ausnutzung zelleigener Transportsysteme zum Transfer von Nukleinsäuren durch die Kernhülle. Patent-Offenlegungsschrift DE 19933939 A1, German Patent and Trade Mark Office, Munich, Germany.
17. Rothmann-Cosic, K., Wessendorf, H., Helfrich, J., Thiel, C., Riemen, G., Brosterhus, H., Müller-Hartmann, H., Weigel, M., Lorbach, E., Nix, M., and Siebenkotten, G. (2002) Buffer solution for electroporation and a method comprising the use of the same. Europäische Patentschrift EP 1390518 B1, European Patent Office, Munich, Germany.
18. Pontén, J., and Macintyre, E. H. (1968) Long term culture of normal and neoplastic human glia. *Acta Pathol Microbiol Scand* **74**, 465–486.
19. Beckman, G., Beckman, L., Pontén, J., and Westermark, B. (1971) G-6-PD and PGM phenotypes of 16 continuous human tumor cell lines. Evidence against cross-contamination and contamination by HeLa cells. *Hum Hered* **21**, 238–241.
20. Westermark, B., Pontén, J., and Hugosson, R. (1973) Determination for the establishment of permanent tissue culture from human gliomas. *Acta Path Microbiol Scand* **81**, 791–805.
21. Bigner, D. D., Bigner, S. H., Pontén, J., Westermark, B., Mahaley, M. S., Ruoslahti, E., Herschman, H., Eng, L. F., and Wikstrand, C. J. (1981) Heterogeneity of genotypic and phenotypic characteristics of fifteen permanent cell lines derived from human gliomas. *J Neuropathol Exp Neurol* **40**, 201–229.
22. Akslen, L. A., Andersen, K. J., and Bjerkvig, R. (1988) Characteristics of human and rat glioma cells grown in a defined medium. *Anticancer Res* **8**, 797–803.
23. Gracia, E., Fischer, U., ElKahloun, A., Trent, J. M., Meese, E., and Meltzer, P. S. (1996) Isolation of genes amplified in human cancers by microdissection mediated cDNA capture. *Hum Mol Genet* **5**, 595–600.

Chapter 12

An Efficient Transfection Method for Mouse Embryonic Stem Cells

Jun-Yang Liou, Bor-Sheng Ko, and Tzu-Ching Chang

Abstract

Embryonic stem (ES) cells are an important source of stem cells in tissue engineering and regenerative medicine because of their high self-renewal capacities and differentiation potentials. However, the detailed molecular mechanisms controlling the differentiation and renewal programs in ES cells remained unclear. One of the difficulties in understanding these mechanisms substantially results from the low efficacies of gene manipulation by delivering exogenous gene expression or knockdown of endogenous gene expression with small interfering RNA (siRNA) in ES cells. Here we describe an optimized protocol for efficiently transfecting mouse ES cells by Effectene, a liposome-based method. The high transfection efficiency in mouse ES cells is demonstrated in this chapter by (1) achieving a percentage of enhanced green fluorescence protein (EGFP) expression in >98% embryoid bodies after introducing plasmids encoding the protein and (2) decreased SOX-2 and Oct-3/4 expression and subsequent morphological evidence of cell differentiation after introducing siRNA expression for suppressing SOX-2 and Oct-3/4, which are known to be essential for maintenance of stem cell properties in mouse ES cells.

Key words: Mouse embryonic stem cells, transfection, efficiency, green fluorescent protein, small interfering RNA.

1. Introduction

Mouse embryonic stem (mES) cells are derived from the inner mass of embryos in the blastocyst stage of development, and they have been shown to have extensive self-renewal capacities and differentiation abilities for all the three germ layers including ectoderm, mesoderm, and endoderm (1, 2). With their high proliferation potentials, mES cells can provide unlimited cell sources for the study of developmental biology, tissue engineering, regenerative medicine, and cell-based gene therapy. The

B. Zhang, E.J. Stellwag (eds.), *RNAi and MicroRNA-Mediated Gene Regulation in Stem Cells*, Methods in Molecular Biology 650, DOI 10.1007/978-1-60761-769-3_12, © Springer Science+Business Media, LLC 2010

possible operating and regulatory genetic programs for mES cell differentiation and proliferation are therefore of intense scientific interest and merit thorough investigation. However, these studies are not easily conducted, partly because of the inadequate methods for genetic manipulation by transfection. Although viral vector-based transfection methods had been used in mES cells with some limited success, nonviral gene delivery methods such as liposome, electroporation, and nucleofection are emerging approaches because of the safety concerns related to viral transfection and even as viable alternatives to viral transfection. Effectene and lipofectamine, for example, had been reported to have 20–70% (3) and 50–80% (4) transfection efficiency in mES cells. A rate of 63–66% for mES cells transfection efficiency had been reported by nucleofection, in contrast of 6.41% by electroporation (5).

Alteration of gene expression by small interfering RNA (siRNA) has been described as a powerful gene knockdown approach in studying molecular and cellular biology in eukaryotic systems. Recently, this approach has been applied to mES cells (6–8). But the success of the siRNA knockdown method is dependent on an efficient protocol for delivery of DNA to mES cells.

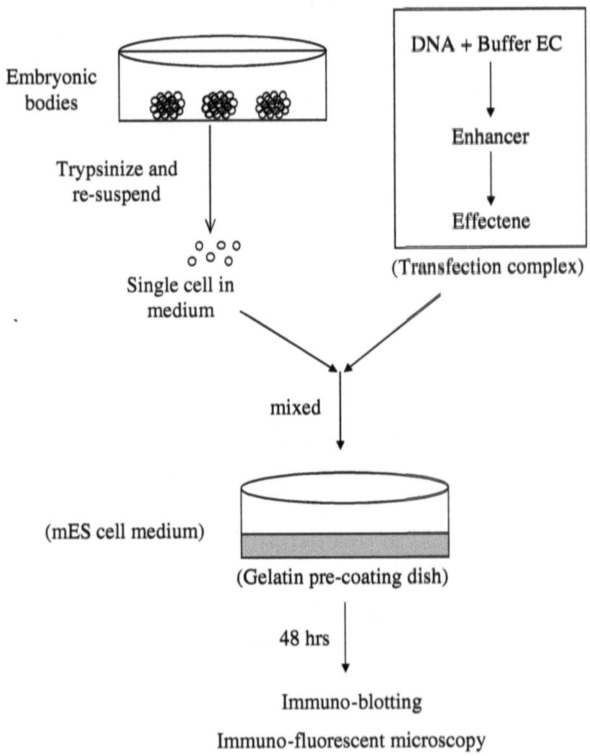

Fig. 12.1. A schematic illustration of the transfection method for mouse ES cells. Detailed information and protocol are described in this chapter.

Consequently, the search for high transfection-efficiency methods to use with siRNA is essential. We have recently reported an optimized protocol for gene transfection in mES cells (9). A high transfection efficiency could be observed with either exogenously forced-expressed genes, such as EGFP, or delivery of siRNA. The detailed protocol (schematic illustration in **Fig. 12.1**) and representative results will be described below.

2. Materials

2.1. Gelatin Coating of Tissue Culture Vessels

1. 2% Gelatin solution Type B: From bovine skin (Sigma-Aldrich, Inc., Saint Louis, Missouri).

2.2. mES Cell (CCE) Culture

1. CCE is a mouse embryonic stem (ES) cell line derived from 129/Sv mouse strain (provided by StemCell Technologies, Vancouver, BC, Canada).
2. 0.05% Trypsin-EDTA (Gibco/BRL, Invitrogen, Grand Island, NY).
3. Dulbecco's modified Eagle's medium (DMEM) high glucose 1x (with glutamine, Gibco/BRL, Invitrogen).
4. Fetal bovine serum (Biological Industries Ltd., Kibbutz Beit Haemek, Israel).
5. MEM nonessential amino acid 100x solution (MEM NEAA, Gibco/BRL, Invitrogen).
6. Sodium pyruvate, 100 mM (100×) (Gibco/BRL, Invitrogen).
7. Recombinant mouse leukemia inhibitory factor (LIF, Millipore Corporation, Bedford, MA).
8. 2-Mercaptoethanol (2-ME, Gibco/BRL, Invitrogen).
9. The CCE medium contains DMEM supplemented with 15% FBS, 100 U/mL penicillin, 100 μg/mL streptomycin, 0.1 mM MEM NEAA, 1 mM sodium pyruvate, 10 ng/mL LIF, and 0.1 mM 2-ME.

2.3. Transfection Reagents

1. Effectene transfection reagent (contains Buffer EC, Enhancer, and Effectene) (Qiagen, GmbH, Hilden, Germany).
2. pEGFP-C1 plasmid (Clontech Laboratories, Inc. Mountain View, CA).
3. Control (scramble) siRNA (control siRNA is a nontargeting 20–25 nt siRNA designed as a negative control) (Santa Cruz Biotechnology, Santa Cruz, CA).

4. Oct-3/4 siRNA (specific for mouse; Oct-3/4 siRNA for mouse is a pool of 3 target-specific 20–25 nt siRNAs designed to knockdown gene expression) (Santa Cruz Biotechnology).
5. SOX-2 siRNA (specific for mouse; SOX-2 siRNA for mouse is a pool of 3 target-specific 20–25 nt siRNAs designed to knockdown gene expression) (Santa Cruz Biotechnology).

2.4. Immunoblotting

1. RIPA lysis buffer (contains 0.5 M Tris-HCl, pH 7.4, 1.5 M NaCl, 2.5% deoxycholic acid, 10% NP-40, 10 mM EDTA) (Upstate, Lake Placid, NY).
2. Protease inhibitor cocktail (contains Aprotinin, Bestatin, Calpain inhibitor I, Calpain inhibitor II, Chymostatin, E-64, Leupeptin, α2-Macroglobulin, PefablocSC, Pepstatin, PMSF, TLCK-HCl, Trypsin inhibitor) (Roche Diagnostics GmbH, Mannheim, Germany).
3. Bio-Rad Protein Assay kit (Bio-Rad laboratories Inc., Hercules, CA).
4. Laemmli sample buffer (contains 62.5 mM Tris-HCl, pH 6.8, 25% glycerol, 2%SDS, 0.01% bromophenol blue) (Bio-Rad laboratories Inc.).
5. PVDF membrane (Immobilon-P) (Millipore Corporation, Bedford, MA).
6. Monoclonal antibody against mouse Oct-3/4 (Santa Cruz Biotechnology).
7. Rabbit polyclonal antibody against GFP (Santa Cruz Biotechnology).
8. Mouse monoclonal antibody against actin (Sigma-Aldrich).
9. Rabbit polyclonal antibody against mouse SOX-2 (Chemicon International Inc., Temecula, CA).
10. Horseradish peroxidase-conjugated secondary antibody (Chemicon International Inc.).
11. Enhanced chemiluminescence (ECL reagent, PerkinElmer, Bridgeport Avenue Shelton, CT).

3. Methods

3.1. Gelatin Coating of Tissue Culture Vessels (see Note 1)

1. Dispense sufficient gelatin solution into a culture vessel to cover the bottom completely. Suggested volumes are 3 mL per T25-cm^2 flask or 60-mm tissue culture dish; 7–8 mL per 100-mm tissue culture dish.

2. Incubation of gelatin solution in culture vessels for at least 10 min at room temperature.
3. Aspirate gelatin solution and keep the container lid open in the hood until completely dry (*see* **Note 2**).

3.2. Passage and Maintenance of mES Cells: (see Note 3)

1. CCE cells were maintained in complete medium in a suitable size of culture vessel for 48 h, and CCE cell colonies (also defined as embryoid bodies, EB) can be readily observed.
2. Aspirate all media from the culture vessel.
3. Rinse cultured CCE cells once with phosphate-buffered saline (PBS).
4. Add sufficient prewarmed (at room temperature) trypsin-EDTA solution (0.5 mL per T25-cm^2 flask or 60-mm tissue culture dish and 1 mL per 100-mm tissue culture dish) to cover the cells completely.
5. Incubate at room temperature until the cells begin to detach from the surface of the culture vessel. It usually requires 2 or 3 min for most ES cell lines cultured on gelatinized dishes.
6. Harvest detached cells from culture vessel, pipetting 5 times and transfer into a tube containing a solution of DMEM 10% FBS; centrifuge the cell suspension to make cells form a pellet (240 ×*g* for approximately 10 min) (*see* **Note 4**).
7. Aspirate the medium and resuspend the cells gently in approximately 2 mL of CCE medium. Pipette up and down against the bottom of the tube 4–6 times to ensure cell pellet is disrupted to a single cell suspension (*see* **Note 5**).
8. If doing a 1/10 split of the cells (*see* **Note 6**), transfer 0.2 mL of cells onto a freshly prepared gelatinized dish containing the appropriate volume of CCE medium.
9. Continue to passage (usually every other day) and maintain CCE cells in an undifferentiated state.

3.3. Plasmid DNA or siRNA Transfection of ES Cells

1. The following protocol is for transfection of CCE cells in 60-mm dishes with Effectene transfection reagent.
2. At the beginning of transfection, adherent CCE cells should be prepared as a cell suspension in the appropriate medium (*see* **Note 7**). Preparing suspended ES cells is conducted according to steps 1–5 in **Section 3.2** described above.
3. Aspirate the medium and resuspend CCE cells in approximately 5 mL of CCE medium. And then make the optimal

split of the cells into different tubes on the basis of the following different plasmid transfections (*see* **Note 8**).

4. Begin to prepare the transfection reagent mixture while CCE cells are being centrifuged.
5. Dilute 2 μg DNA or 2 μg siRNA (*see* **Notes 9 and 10**) that was dissolved in nuclease-free TE buffer with the DNA-condensation buffer, Buffer EC, to a total volume of 150 μL.
6. Add 6.4 μL Enhancer and mix by vortexing for 1 s (*see* **Note 11**).
7. Incubate at room temperature (15–25°C) for 2–5 min; then centrifuge the solution at 1,800*g* for 10 s to remove drops from the top of the tube.
8. Add 16 μL Effectene Transfection Reagent (*see* **Note 12**) to the DNA-Enhancer mixture. Mix by vortexing for 10 s.
9. Incubate the samples for 5–10 min at room temperature (15–25°C) to allow transfection complex formation and centrifuge the solution at 1,800*g* for 10 s to remove drops from the top of the tube.
10. Add the above transfection complex to the tubes containing the suspended ES cells, which were prepared as described above. Gently mix by pipetting up and down twice and immediately add the cell-transfection suspension into the 60-mm dishes with ES cell medium. Gently swirl the dish to ensure uniform distribution of the cell-transfection suspension.
11. To avoid possible cytotoxicity induced by the transfection reagents, remove the Effectene–DNA complexes after 24 h by washing the cells once with PBS, and adding 5 mL fresh medium.
12. Incubate the cells with the transfection suspension under their normal growth conditions for an appropriate time for expression of the transfected gene. The incubation time is determined by the assay and gene used.
13. Monolayer and differentiated morphology of CCE cells transfected with siRNA (SOX-2 or Oct-3/4) can be examined using a phase-contrast microscope (Olympus CKX41) (**Fig. 12.2**).

3.4. Immunoblotting Assay

1. Transfected CCE cells are washed with PBS and lysed in ice-cold RIPA lysis buffer (Upstate) containing protease inhibitor cocktail (Roche).
2. The lysates are clarified by centrifugation at 20,800×*g* for 5 min at 4°C. The supernatants are collected and protein

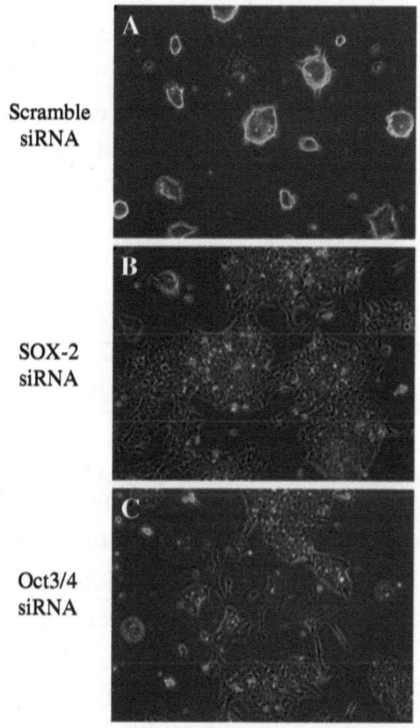

Fig. 12.2. Monolayer, and differentiated morphology of CCE cells transfected with siRNA of SOX-2 or Oct-3/4 were examined by phase-contrast microscopy. **(A)** Transfection with scrambled siRNA. **(B)** Transfection with SOX-2 siRNA. **(C)** Transfection with Oct-3/4 siRNA.

concentrations are determined using a Bio-Rad Protein Assay kit.

3. Protein extracts (30 μg) are mixed with equal volume of Laemmli sample buffer (Bio-Rad) and incubated at 95°C for 15 min. Protein mixtures are resolved on an SDS-8% polyacrylamide gel electrophoresis (SDS-PAGE) and transferred to PVDF membrane (Millipore) (100 V, 1 h).
4. Membranes are blocked with 5% blocking grade skim-milk in PBS-T (0.1% Tween 20) for 1 h and subsequently incubated with first antibody, rabbit polyclonal antibody against mouse SOX-2, diluted at 1:3000; mouse monoclonal antibody against mouse Oct-3/4 at 1:10,000, rabbit polyclonal antibody against GFP at 1:3000, and goat polyclonal antibody against actin at 1:5000 at 4°C for overnight.
5. The membrane is washed with PBS-T three times and incubated with horseradish peroxidase-conjugated secondary antibody (Santa Cruz) at room temperature for 1 h.
6. Subsequently, the membrane is washed three times with PBS-T and analyzed by enhanced chemiluminescence (PerkinElmer).

3.5. Fluorescent Microscopy

1. CCE cells transfected with CMV promoter-driven pEGFP overexpression vector for 48 h form embryonic bodies (EB).
2. Cells were examined by fluorescent microscope (Leica DMIRB) and the percentage of GFP-positive EB was calculated (number of GFP positive EB/number of total EB).

4. Notes

1. CCE mES cells can be maintained in the medium under feeder cell-free conditions. However, it is essential to precoat the bottom of culture vessels with 0.2% gelatin.
2. Gelatinized dishes or flasks can be stored (remain sterile) at room temperature or at 4°C for at least 2 weeks.
3. The passage of CCE cells needs to be done before the cells reach confluence and prior to the growth medium becoming acidic. The density of the CCE colonies (EB) should not exceed 50–70% of the culture vessel surface. The frequency of passage is dependent upon the growth rate of the cells, but CCE cell lines require passage every second day (usually no less frequently than 48–72 h). It is critical that the passage needs to be done regularly since overgrowth of the cultures promotes the differentiation of the CCE cells.
4. The serum is present in this wash step to inactivate residual trypsin activity.
5. It is absolutely essential to ensure a single cell suspension is achieved as transfer of cell clumps will promote differentiation and decrease transfection efficiency.
6. CCE cells do not grow well if they are plated at a low density. On the other hand, if the cell density is too high, differentiation is promoted. It is best to achieve a culture density where adherent cells cover approximately 40–50% of the surface area of the culture vessel at 24 h after plating. Absolute numbers of cells transferred to the various sizes of culture dishes are dependent upon the ES cell line used, approximately 5×10^5 cells per 100-mm dish and 2×10^5 per 25-cm^2 flask or 60-mm dish.
7. The separated, single cells are much easier to transfect than round embryonic bodies.
8. The optimal number of cells for transfection is 2×10^5 per 25-cm^2 flask or 60-mm dish and 5×10^5 cells per 100-mm dish.

9. Plasmid DNA quality significantly influences several transfection parameters such as efficiency, reproducibility, and toxicity, as well as interpretation of results. Therefore only plasmid DNA of the highest purity should be used.
10. The optimal quantity of plasmid DNA used for transfection is determined by the properties of the transfected plasmid, and includes the type of promoter, origin of replication, and plasmid size. Toxic effects may arise if transfected with an excessively high amount of DNA or RNA/Effectene.
11. Always keep the ratio of DNA to Enhancer solution constant. In CCE cells, the optimal ratio of DNA (μg) to Enhancer (μL) is typically in the range between 1:1.6 or 1:3.2.
12. Always keep the ratio of DNA to Effectene constant. In ES cells, the optimal ratio of DNA (μg) to Effectene (μL) is typically in the range between 1:4 or 1:8.

Acknowledgments

This work was supported by grants from Taiwan, National Health Research Institutes (NHRI-97A1-CVPP03-014) and National Science Council (NSC97-3111-B-400-004).

References

1. Evans, M. J., Kaufman, M. H. (1981) Establishment in culture of pluripotential cells from mouse embryos. *Nature* **292**, 154–156.
2. Martin, G. R. (1981) Isolation of a pluripotent cell line from early mouse embryos cultured in medium conditioned by teratocarcinoma stem cells. *Proc Natl Acad Sci USA* **78**, 7634–7638.
3. Bugeon, L., Syed, N., Dallman, M. J. (2000) A fast and efficient method for transiently transfecting ES cells: Application to the development of system for conditional gene expression. *Transgenic Res* **9**, 229–232.
4. Ward, C. M., Stern, P. L. (2002) The human cytomegalovirus immediate early promoter is transcriptionally active in undifferentiated mouse embryonic stem cells. *Stem Cells* **20**, 472–475.
5. Lakshmipathy, U., Pelacho, B., Sido, K., Linehan, J. L., Coucouvanis, E., Kaufman, D. S. et al. (2004) Efficient transfection of embryonic and adult stem cells. *Stem Cells* **22**, 531–543.
6. Schaniel, C., Li, F., Schafer, X., Moore, T., Lemischka, I.R., Paddison, P. J. (2006) Delivery of short hairpin RNAs-triggers of gene silencing-into mouse embryonic stem cells. *Nat Methods* **3**, 397–400.
7. Hough, S. R., Clements, I., Welch, P. J., Wiederholt, K. A. (2006) Differentiation of mouse embryonic stem cells after RNA interference-mediated silencing of OCT4 and Nanog. *Stem Cells* **24**, 1467–1475.
8. Chen, S., Choo, A., Wang, N. D., Too, H. P., Oh, S. K. (2007) Establishing efficient siRNA knockdown in mouse embryonic stem cells. *Biotechnol Lett* **29**, 261–265.
9. Ko, B.S., Chang, T.C., Shyue, S. K., Chen, Y.C., Liou, J.Y. (2009) An efficient transfection method for mouse embryonic stem cells. *Gene Therapy* **16**, 154–158.

Chapter 13

Semi-quantitative Analysis of Transient Single-Cell Gene Expression in Embryonic Stem Cells by Femtoinjection

Mikako Saito and Hideaki Matsuoka

Abstract

Single-cell manipulation supporting robot (SMSR) has enabled femtoinjection, a high-throughput and semi-quantitative microinjection in the range of femtogram (fg) DNA and other molecules. An enhanced green fluorescent protein (EGFP) gene expression vector can be introduced directly into mouse embryonic stem cells at 100 cells/h with a 10% success rate. The intensity of EGFP fluorescence in single ES cells or single colonies of ES cells increases as the concentration of an EGFP gene expression vector in the injection capillary increases from 1 to 50 ng/μL. On the other hand, the knockdown of EGFP gene expression can be demonstrated by femtoinjection of siRNA against EGFP or an shRNA expression vector using an EGFP expressing ES cell line. Femotoinjection can provide a useful method for quantitative analysis of transient gene expression in single cells using RNAi.

Key words: Femtoinjection, single-cell manipulation supporting robot, embryonic stem cell, gene expression vector, RNAi, transient gene expression, single-cell analysis.

1. Introduction

In order to use embryonic stem (ES) cells in regenerative engineering, it is essential to elucidate the mechanisms that regulate the maintenance of the undifferentiated state and the induction to specific differentiated states. A number of studies have been undertaken to examine the roles of factors such as *Stat3* (1), *Oct3/4* (2, 3), *Sox2* (4), and *Nanog* (5). At present, however, it is unclear how these factors interact, for example, whether they are additive, cooperative, or antagonistic. Moreover, some genes, such as *Oct3/4*, show variations in the levels of their expression during ES cell differentiation (2). Microinjection offers

B. Zhang, E.J. Stellwag (eds.), *RNAi and MicroRNA-Mediated Gene Regulation in Stem Cells*, Methods in Molecular Biology 650, DOI 10.1007/978-1-60761-769-3_13, © Springer Science+Business Media, LLC 2010

a potentially powerful method for the analysis of the dynamics of expression and interaction of multiple genes because it enables real-time regulation of transient expression of multiple genes. Up-regulation of expression can be achieved by microinjection of over-expression vectors, while down-regulation can be accomplished by microinjection of siRNAs or shRNA expression vectors.

Until recently, however, microinjection proved too technically demanding for routine use in gene expression analyses in ES cells. One particular problem is that ES cells offer an extraordinarily difficult target because their diameter may be as little as 15 μm and they have a sticky cell membrane. As a consequence, gene microinjection into ES cells was unsuccessful until an experiment performed in our laboratory in 2005 by a particularly proficient microinjection expert. Even so, the success rate was only 0.2% and, thus, far from a practical level. To develop a higher throughput method, we built a robot that can support the microinjection operator: the single-cell manipulation supporting robot (SMSR). The basic idea behind the SMSR is that the operator is able to concentrate on the actual microinjection, while the robot performs the remaining operational steps (6, 7). In 2007, SMSR enabled high-throughput injection (100 cells/h), a high success rate of up to 10%, and semi-quantitative injection in the range of fg, even by ordinary students (8). Therefore, this technology may be called femtoinjection in contrast to conventional microinjection. By following the operational procedures described in this chapter, femtoinjection of gene expression vectors, RNAi, and various combinations of these vectors is possible. Enhanced green fluorescent protein (EGFP) and small heterodimer partner (SHP) (9, 10) are used as examples of target genes in the demonstrative experiments.

2. Materials

2.1. Plasmids and Restriction Enzymes

1. EGFP gene expression vector (pCAG-EGFP, 7.3 kb) (Clontech Laboratories, Palo Alto, CA).
2. Red fluorescent protein (DsRed) gene expression vector (pCAG-IRES-DsRed, 7.8 kb) (*see* **Note 1**).
3. cHA (cDNA of hemagglutinin)-EGFP gene expression vector (pCAG-cHA-IP-EGFP), in which IP is a construct of the internal ribosomal entry site (IRES) and puromycin-tolerance gene (*see* **Note 1**).
4. Restriction enzymes, ScaI and BamHI.

2.2. Cells and Media

1. EB3 cells, a clone of feeder-free mouse ES cells (*see* **Note 1**).
2. G4-2 cells, an EGFP-expressing stable clone of mouse ES cells developed from EB3 (*see* **Note 1**).
3. SHP-G, an SHP-EGFP-expressing stable clone of mouse ES cells (*see* **Note 2**).
4. Medium for EB3, G4-2, and SHP-G cells is GMEM supplemented with 10% fetal calf serum (FCS), 1 mM sodium pyruvate, 10^{-4} M 2-mercaptoethanol, 1× nonessential amino acids (NEAA), and 1,000 U/mL of leukemia inhibitory factor (LIF) on gelatin-coated polystyrene culture dishes 35 mm^{ϕ} (diameter).

2.3. Materials for Femtoinjection

1. Glass capillary (1 mm^{ϕ}) (#BF100-78-10, Sutter Instrument Co., Novato, CA, USA)
2. Coordinate standard chip (16 mm × 6 mm) (*see* **Note 3**) (11)
3. Culture dishes 35 mm^{ϕ} (diameter)
4. Polyethylene (PE) tube (1.67 mm^{ϕ}) (Hibiki-No.5, KK-Kunii, Tokyo, Japan)

2.4. Fluorescent Dye

1. SYTO 83 nucleic acid stain.

3. Methods

Prepare gene expression vectors and RNAi according to standard protocols. Adjust the concentrations of the vectors and RNAi to the required levels following preliminary microinjection experiments. Since it is difficult to determine the quantity actually injected into a target single cell, obtain an estimate of the injected quantity by ejection experiments using micro water drops. The ejected quantity varies with the capillary tip diameter, ejection pressure, the pressure holding time, and the vector concentration in the capillary. Overall, we suggest that the best approach to quantitative microinjection is to regulate the concentration in the injection capillary.

3.1. Preparation of Gene Expression Vectors

1. Prepare pCAG-SHP-IP-EGFP (**Fig. 13.1B** from pCAG-cHA-IP-EGFP (**Fig. 13.1A**) by replacing cHA with SHP (full-length cDNA: 1,127 bp) cloned separately by RT-PCR as described in **Note 4**.
2. Add 100 μL of a pCAG-SHP-IP-EGFP (ca. 20 μg in sterilized pure water, *see* **Note 5**) solution to a ScaI solution

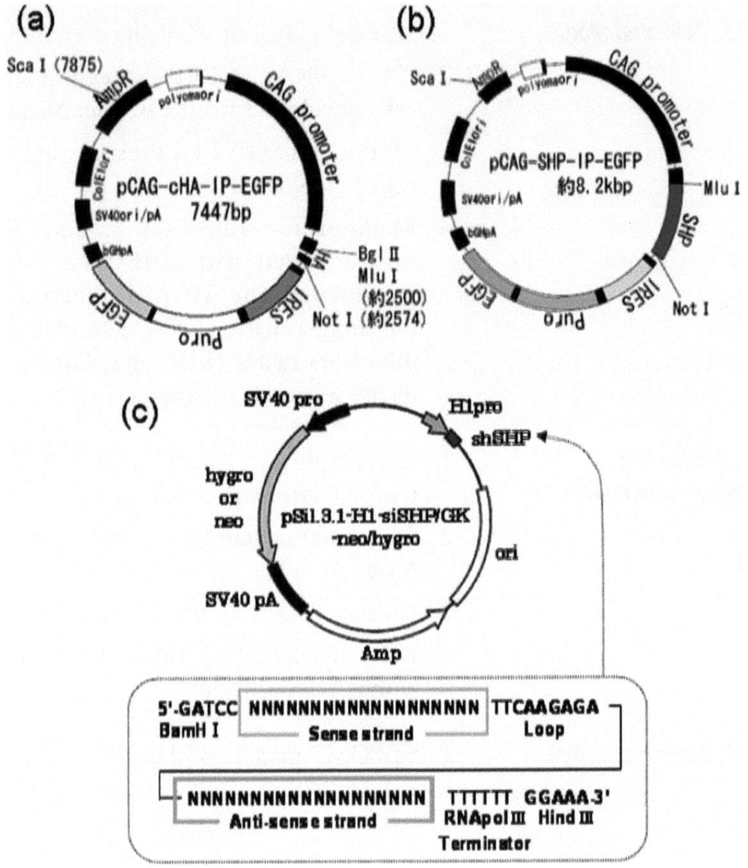

Fig. 13.1. Gene expression vectors used in this chapter. (**A**) pCAG-cHA-IP-EGFP, (**B**) pCAG-SHP-IP-EGFP, (**C**) p-shSHP.

(20 μL/DNA 2 μg) and incubate for 16 h at 37°C to linearize the vector. To this reaction solution, add 100 μL phenol (neutralized with Tris)-chloroform (1:1). Mix thoroughly, centrifuge at 17,300*g* for 10 min, and then transfer the supernatant to a new microtube (1.5 mL). To this microtube, add sodium acetate (10% v/v of supernatant, pH 5.3) and 250μL ethanol. Mix thoroughly and place in a −80°C freezer for 20 min. Centrifuge at 17,300*g* for 15 min and remove the supernatant. Rinse and resuspend the precipitate in 70% ethanol, centrifuge at 17,300*g* for 5 min, remove the supernatant, and dry the precipitate. Dissolve the dried precipitate in 100 μL sterilized pure water. Purify the DNA by electrophoresis on a 1% agar gel.

3. Linearize the pCAG-EGFP and pCAG-IRES-DsRed vectors as described for pCAG-SHP-IP-EGFP, except for use of the restriction enzymes, ScaI and BamHI, respectively.
4. Adjust DNA concentrations to the required values in the range from 1 to 50 ng/μL.

3.2. Preparation of siRNAs and shRNA Expression Vectors

1. Design siRNAs against EGFP, DsRed, and SHP (siEGFP, siDsRed, and siSHP) according to the guidelines proposed by Ui-Tei et al. (12). Among candidate siRNAs, select one with the highest activity against the target gene by lipofection (*see* **Note 6**). The followings are siRNAs used in our experiments.
   siEGFP
   sense strand: 5′-GCAAGCUGACCCUGAAGUUTT-3′
   antisense strand: 5′-AAUUUCAGGGUCAGCUUGCTT-3′
   siDsRed
   sense strand: 5′-GCAAGCUGACCCUGAAGUUTT-3′
   antisense strand: 5′-AACUUCAGGGUCAGUUUGCTT-3′
   siSHP
   sense strand: 5′-GGAACAAGAUACUAACCAUTT-3′
   antisense strand: 5′-AUGGUUAGUAUCUUGUUCCTT-3′
2. Design an shRNA against SHP (shSHP) using the sequence of siSHP. Insert the shSHP and a loop insert into pSilencer3.1-H1-neo to construct the shSHP expression vector (p-shSHP) (**Fig. 13.1C**).
3. Linearize the p-shSHP as described above (**Section 3.1** and step 2).
4. Adjust the DNA concentrations to the required values in the range from 1 to 50 ng/μL.

3.3. Preparation of a Capillary for Femtoinjection

1. Put 30–40 glass capillaries into a test tube (16 mm$^{\phi}$). Fill the test tube with methanol and shake gently to wash the glass capillaries. Repeat the methanol wash, and then rinse the capillaries with pure water (*see* **Note 5**). Remove the water, cover the test tube with aluminum foil, and place in a dryer at 180°C for 3 h.
2. Set the operational parameters of the laser puller (Model P2000, Sutter Instrument Co.) at the following values: Line 1 (Heat = 320, Fil = 5, Vel = 12, Del = 220, Pul = 100), Line 2 (Heat = 320, Fil = 8, Vel = 8, Del = 184, Pul = 150), Line 3 (Heat = 320, Fil= 15, Vel = 15, Del = 136, Pul = 215). Using these conditions, pull a glass capillary and check its shape under a microscope. The tip diameter should be 0.6–0.8 μm and the length of the thin neck should be 600 μm. Discard capillaries with an unusual shape or that are broken. If possible, check the tip diameter with a scanning electron microscope (SEM). If capillaries with the requisite size and shape are not produced under these conditions, alter the parameters described above to identify appropriate conditions.

3. Cut a PE tube (1.67 mm$^{\phi}$) into 3–4 cm lengths. Pinch the ends of each piece of tubing and heat its midsection with an alcohol lamp until it almost melts. At this point, stretch the tubing to form a thin stringy section. Cut the stretched tube at its midpoint to make a pair of PE stringy tubes.
4. Insert a syringe needle (**Fig. 13.2ii**) into the thick end of a PE stringy tube (**Fig. 13.2iii**). Dip the thin end of the stringy tube into a solution containing a vector or RNAi and suck the solution into the tube. Pick up the thin end with tweezers and insert it deep into a glass capillary. Eject the solution into the glass capillary.

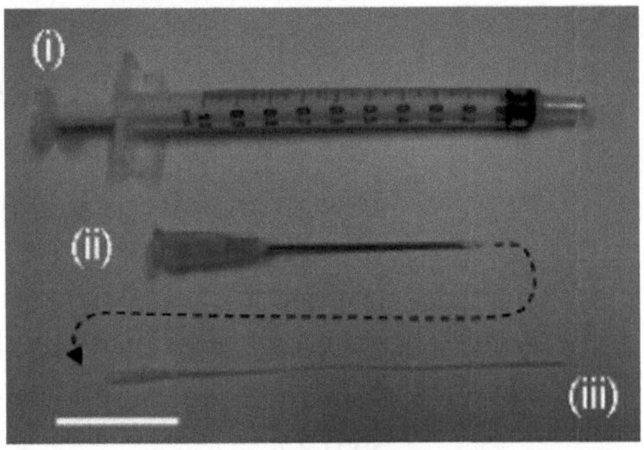

Fig. 13.2. Connecting the PE tube to the syringe needle. **(i)** PE syringe, **(ii)** needle, **(iii)** PE tube. Bar = 2 cm.

5. Check the capillary carefully to ensure that the solution does not contain any air bubbles. If an air bubble is present, hold the capillary vertically with its tip down and flick it gently. Protect the capillary tip from drying by storing the capillaries in a sealed container that maintains humidity.

3.4. Femtoinjection Using SMSR

1. The operational protocol for SMSR is described below. SMSR is commercially available (*see* **Note 7**), With regard to microinjection using other apparatus, also *see* **Note 7**.
2. Attach a coordinate standard (CS) chip to the outer side of the bottom of the culture dish (**Fig. 13.3**) (11). Coat the inside surface of the dish bottom with gelatin. Add a cell suspension containing 5,000–10,000 cells to the culture dish and incubate at 37°C under 5% CO_2 for 12–16 h. During culture, the cells adhere to the dish bottom.
3. Place the culture dish in the dish holder on the automatic stage of the microscope (**Fig. 13.4i, ii**). Position the automatic stage such that the P-point of the CS chip is in the view center (**Fig. 13.3D**). Click the switch to register the

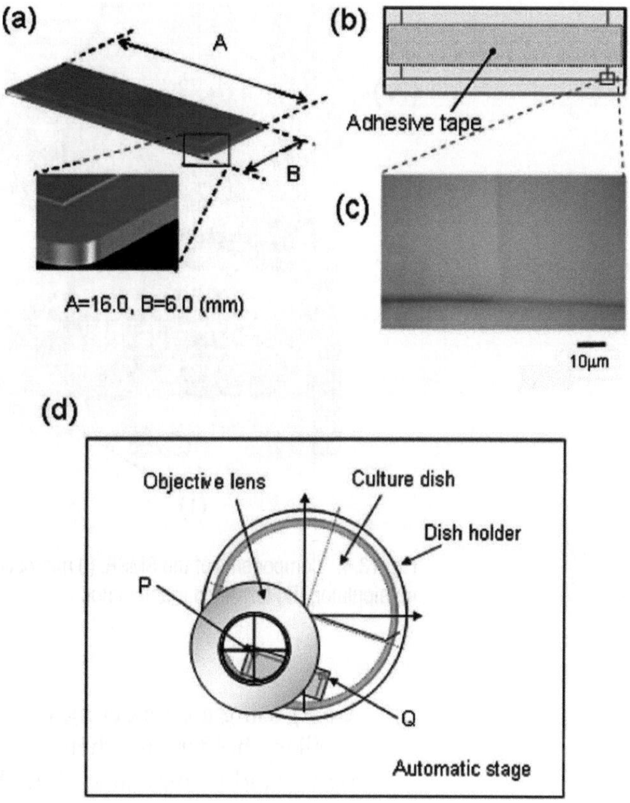

Fig. 13.3. Surface structure and size of a CS chip. (**a**) Three sizes are available; A×B (mm) = 16×6, 16×10, and 40×23; thickness = 0.3 mm. (**b**) Adhesive tape is attached to the top surface of the chip. (**c**) Microscopic view of the edge lines present on the top surface. (**d**) A culture dish with a CS chip on a microscope stage; points P and Q are brought in sequence into the view center to register the origin and the X-axis. The view through the microscope (central area of the objective lens) is illustrated, but is not to scale.

P-point as the origin. Move the automatic stage such that the Q-point of the CS chip is now in the view center. Click the switch to register the line PQ as the X axis. The Y axis is defined automatically as a line that crosses the X axis at a right angle at point P.

4. Display the microscopic image of an arbitrary area of the culture dish on a TV monitor. Select healthy-looking cells in this area and register their XY address using the mouse cursor. Register 50–100 cells per dish. Unless the microinjection is to be performed immediately, return the culture dish to the incubator. In the latter case, repeat step 3 before commencing the injection experiment.
5. Connect the capillary holder to the Pneumatic Pico Pump (PV800, World Precision Instruments Inc., Sarasota, FL,

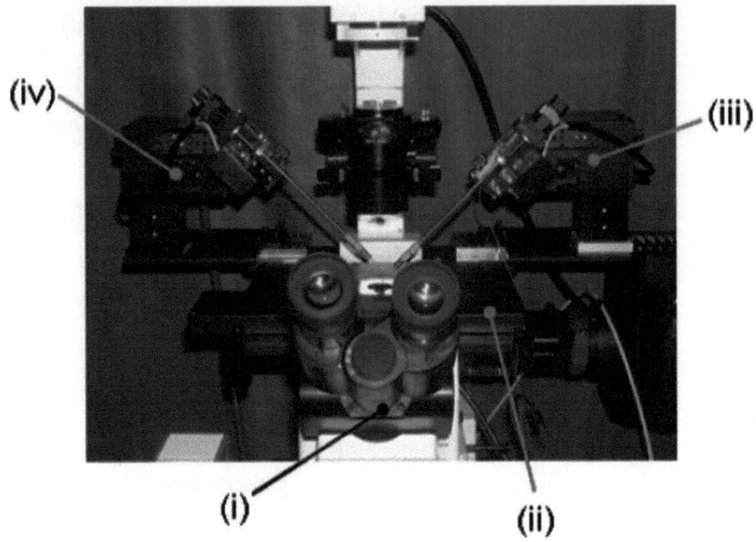

Fig. 13.4. Components of the SMSR. (i) microscope, (ii) automatic stage, (iii) right-hand manipulator, (iv) left-hand manipulator.

USA) using a connecting tube. Place a glass capillary in the capillary holder and then attach the capillary holder to the right-hand manipulator (**Fig. 13.4iii**).

6. Display the list of registered cells on the TV monitor and select a cell by clicking the cell number button. The automatic stage will move to bring the selected cell into the microscope view center. Position the capillary tip just above the cell using the right-hand controller (a track ball for XY-driving and a wheel for Z-driving). Insert the capillary tip through the cell surface and switch on the Pneumatic Pico Pump to eject the solution at 0.70 kgf/cm$^2 \times$30 ms. After ejection, the capillary tip is automatically withdrawn from the cell.
7. Repeat the selection and microinjection steps for all the registered cells in the culture dish. The capillary should be replaced after 5–10 injections.
8. Once all the registered cells in this culture dish have been treated, steps 3–7 can then be applied to the next culture dish.

3.5. Estimation of Quantity of Vector Ejected from the Capillary

1. Prepare many glass capillaries (80–100) with a range of tip sizes by varying the operational parameters of the laser puller (*see* **Section 3.3** and step 2) such that tip diameters are distributed uniformly across the range 0.3–0.9 μm.
2. Prepare a glass capillary with a tip diameter of 2–3 μm for use in preparing water droplets.

3. Cover a glass slide with Parafilm® (Pechiney Plastic Packaging, Inc., Chicago, IL). Under the microscope, place water droplets of 3.5-mm diameter (10.5 μL) on the Parafilm by careful ejection from the 2–3 μm$^{\phi}$ capillary. Insert a 0.3–0.9 μm$^{\phi}$ glass capillary containing 10, 50, or 100 ng/μL pCAG-SHP-IP-EGFP into the water droplet and apply a pulsing pressure of 0.70 kgf/cm^2 × 30 ms to the injector to eject the DNA solution.
4. Measure the DNA concentration in each water droplet by real-time PCR and the manufacturer's standard protocol. Mix the water droplet (10.5 μL) with 12.5 μL of 2×SYBR Green PCR Master Mix, and 1 μL each of the forward and reverse primers for pCAG-SHP-IP-EGFP. Incubate this solution (total volume 25 μL) at 50°C for 2 min and 95°C for 10 min, followed by 50 cycles at 95°C for 15 s and 60°C for 1 min. Calculate $DNA_{W\text{-}EJ}$ (DNA concentration in the water drop after ejection) from a standard curve generated beforehand using known quantities of pCAG-SHP-IP-EGFP.
5. After ejection of the vector, wash each capillary and measure the tip diameter by SEM observation. In **Fig. 13.5**, the capillary tip sizes are classified into 3 groups: 0.31–0.50, 0.51–0.70, and 0.71–0.90 μm.

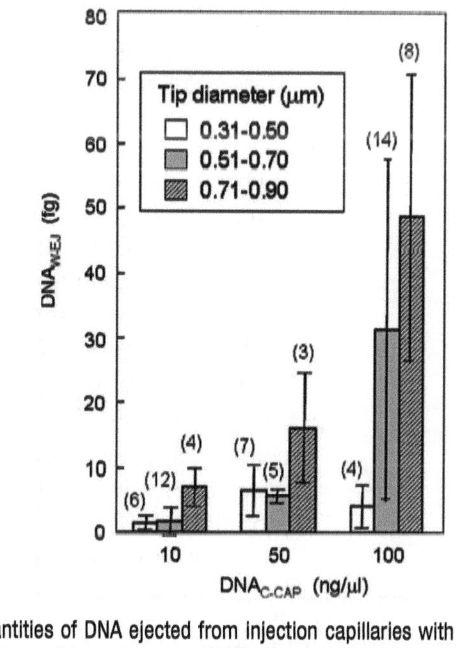

Fig. 13.5. Quantities of DNA ejected from injection capillaries with different tip diameters. The columns show means ± s.d. Figures in parentheses indicate the number of capillaries tested in each case. (Reproduced from Ref. (8) with permission from Springer.)

3.6. Quantitative Injection of EGFP Expression Vector

1. Prepare glass capillaries containing pCAG-EGFP. Adjust the DNA concentration in the capillary ($DNA_{C\text{-}CAP}$) to 1.0–50 ng/μL. Inject the expression vector into ES cells as described in **Section 3.4** and incubate the culture dishes at 37°C under 5% CO_2 for 24 h.
2. Measure the fluorescent intensity produced by the EGFP in each ES cell or cell colony with a fluorescent microscope. Average the fluorescence intensities across the whole area of single cells or single colonies (**Fig. 13.6**) and plot the data versus the injected $DNA_{C\text{-}CAP}$. The results of a typical experiment are shown in **Figs. 13.7 and 13.8**.

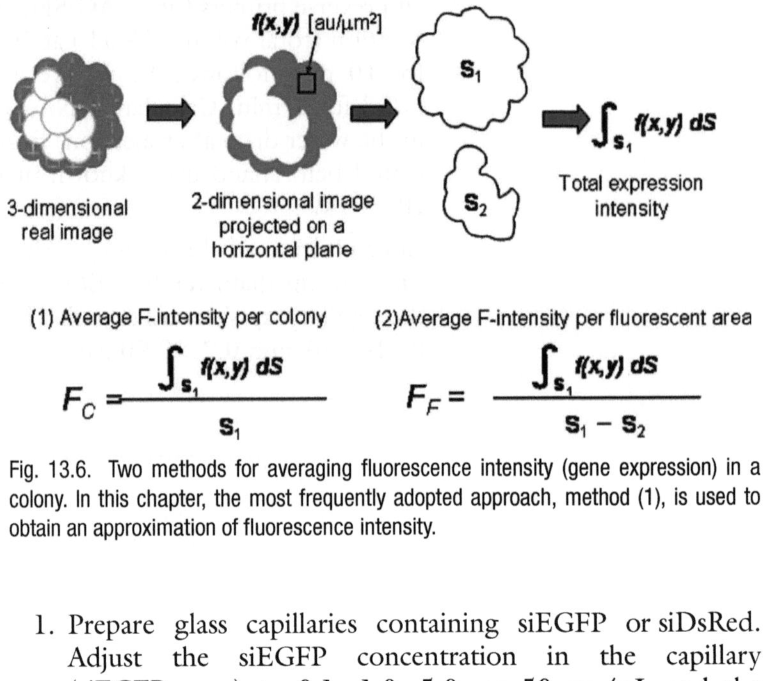

Fig. 13.6. Two methods for averaging fluorescence intensity (gene expression) in a colony. In this chapter, the most frequently adopted approach, method (1), is used to obtain an approximation of fluorescence intensity.

3.7. Quantitative Injection of siRNA

1. Prepare glass capillaries containing siEGFP or siDsRed. Adjust the siEGFP concentration in the capillary ($siEGFP_{C\text{-}CAP}$) to 0.1, 1.0, 5.0, or 50 ng/μL and the siDsRed concentration in the capillary ($siEGFP_{C\text{-}CAP}$) to 50 ng/μL. Inject siEGFP or siDsRed into single G4-2 cells as described in **Section 3.4** and incubate the cells at 37°C under 5% CO_2 for 48 h.
2. Measure the intensity of fluorescence due to EGFP in each colony using a fluorescent microscope. Average the fluorescence intensities across the whole of each single colony and plot against $siRNA_{C\text{-}CAP}$. The results from a typical experiment are shown in **Fig. 13.9**.

3.8. Measurement of Spatial Distribution of RNAi Effect

1. Prepare glass capillaries containing siEGFP and adjust the $siEGFP_{C\text{-}CAP}$ to 50 ng/μL. Inject siEGFP into single G4-2 cells as described in **Section 3.4** and incubate the cells at 37°C under 5% CO_2 for 48 h.

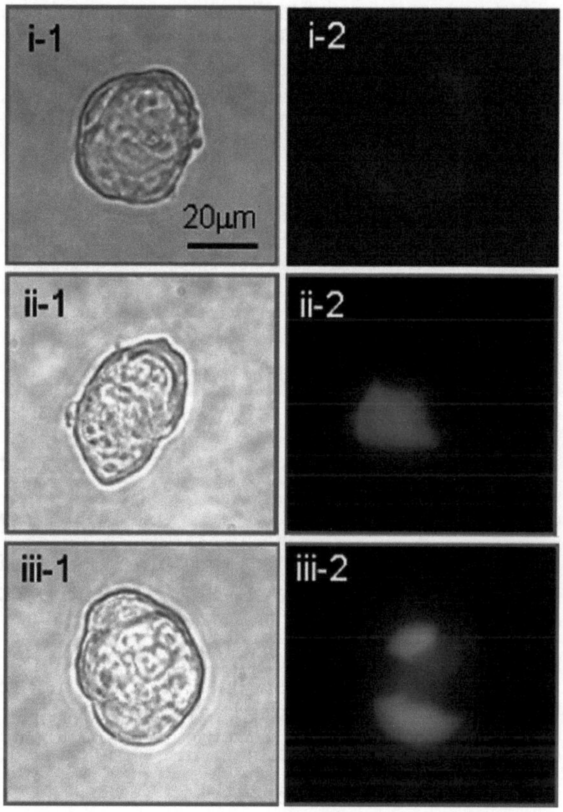

Fig. 13.7. EGFP gene expression observed in EB3 cells after microinjection of different concentrations of the EGFP expression vector. Pressure conditions: 0.70 kgf/cm^2, 30 ms. Estimated concentration of DNA ejected from the capillary: (**i**) 1, (**ii**) 10, and (**iii**) 50 ng/μL. Phase-contrast images (i-1–iii-1) and fluorescent images (i-2–iii-2) were taken 24 h after microinjection. (Reproduced from Ref. (8) with permission from Springer.)

2. Add SYTO 83 (200 nM) to the culture dish to stain the nuclear DNA in ES cells. Incubate the cells at 37°C under 5% CO_2 for 60 min.
3. With a confocal laser-scanning microscope identify single colonies and acquire optical slices of EGFP (Ex/Em = 488/507 nm) and SYTO 83 (Ex/Em = 543/559 nm) fluorescence patterns along the Z-axis. A typical pattern of fluorescence in a colony is shown in **Fig. 13.10**.

3.9. Injection of shRNA Expression Vector into a Target Single Cell of a Colony

1. Culture SHP-G cells at 37°C under 5% CO_2. Select a normal-looking colony and identify a target single cell in the colony. Prepare a glass capillary containing both p-shSHP (50 ng/μL) and p-CAG-IRES-DsRed (50 ng/μL). Inject the vectors into the target cell. After incubation at 37°C under 5% CO_2 for 48 h, analyze the distribution of fluorescence within the colony using a confocal microscope.

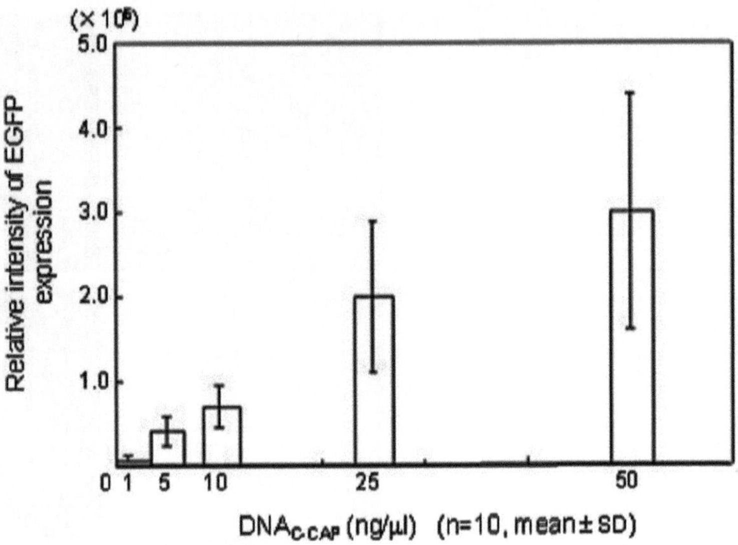

Fig. 13.8. Variation in expression of the EGFP gene after microinjection of different concentrations (1–50 ng/μL) of the EGFP expression vector. (n = 10, mean ± s.d.)

Quenching of EGFP fluorescence due to the knockdown of SHP gene only occurs in cells that show DsRed fluorescence. An example of successful knockdown is shown in **Fig. 13.11**.

4. Notes

1. These vectors and cells were provided by H. Niwa (Center for Developmental Biology, RIKEN, Kobe, Japan) for our demonstration experiments. Make contact with Niwa for their availability.
2. A cell line developed from EB3 by the authors according to the following protocol. Introduce pCAG-SHP-IP-EGFP into EB3 cells by electroporation. Culture EB3 cells for 3–4 days until they become 70–80% confluent in the culture dish. Remove the culture medium and rinse the cells with PBS (0.1 M phosphate-buffer solution, pH 7.4). Add 300 μL trypsin (0.1 %)/EDTA (0.5 mM) to the culture dish to dislodge the cells. After 5 min, add 3 mL of the EB3 culture medium to the dish to stop the trypsin reaction. Collect the cells by centrifugation (175g, 5 min) and rinse them with PBS. Add a cell suspension (1–2×10^7 cells/750 μL PBS) and a solution of pCAG-SHP-IP-EGFP (50 μg DNA/50 μL sterilized pure water) to a microtube. Place the tube on ice for 10 min and then transfer the contents to a cuvette for electroporation. Carry out electroporation using the fol-

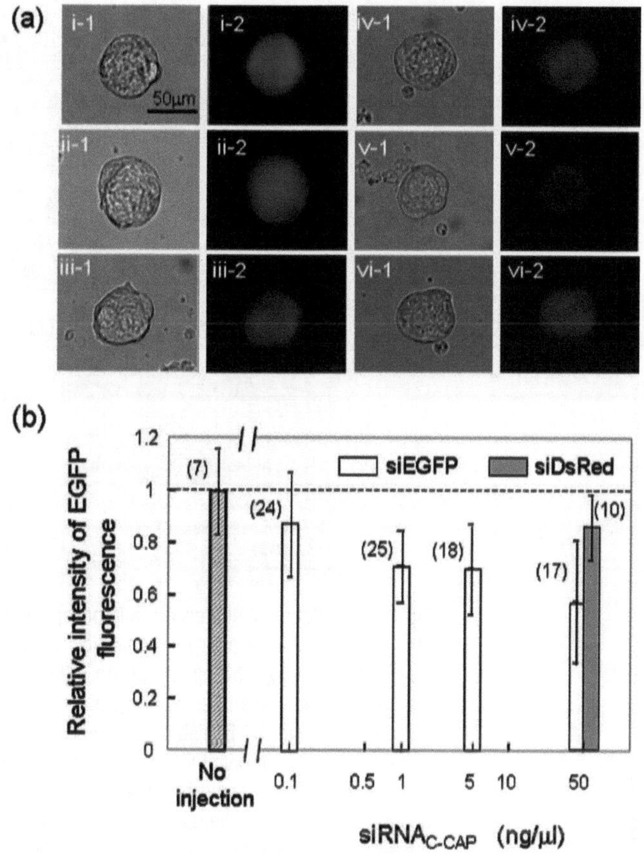

Fig. 13.9. EGFP gene expression in G4-2 cells after microinjection of different concentrations of siRNA. Pressure condition: 0.70 kgf/cm^2, 30 ms. (**a**) Typical images of EGFP expression in ES cell colonies. (*i*) No injection, (*ii*) siEGFP$_{C\text{-}CAP}$ 0.1 ng/μL, (*iii*) siEGFP$_{C\text{-}CAP}$ 1.0 ng/μL, (*iv*) siEGFP$_{C\text{-}CAP}$ 5.0 ng/μL, (*v*) siEGFP$_{C\text{-}CAP}$ 50 ng/μL, (*vi*) siDsRed$_{C\text{-}CAP}$ 50 ng/μL. Phase-contrast images (i-1–vi-1) and fluorescent images (i-2–vi-2) were taken 48 h after injection. (**b**) EGFP gene expression in the different treatment groups. Means ± s.d. are expressed as the values relative to the no injection control. Figures in parentheses indicate the numbers of colonies tested in each group. (Reproduced from Ref. (8) with permission from Springer)

lowing conditions: voltage 0.8 kV, capacitance 3 μF, time constant 0.06–0.08 ms. Cool the cell suspension on ice for 3 min and transfer it to a tube containing 5 mL culture medium. Distribute the suspension among 5 culture dishes (10 cm$^{\phi}$) containing 9 mL culture medium/dish. After 48 h culture, replace the medium by fresh medium containing puromycin (7.5 μg/mL) and incubate the dishes for about 10 days to establish SHP-G.

3. Contact H. Matsuoka (mhide@cc.tuat.ac.jp) about the commercial availability of the coordinate standard chips. Samples of the CS chip will be provided on request.

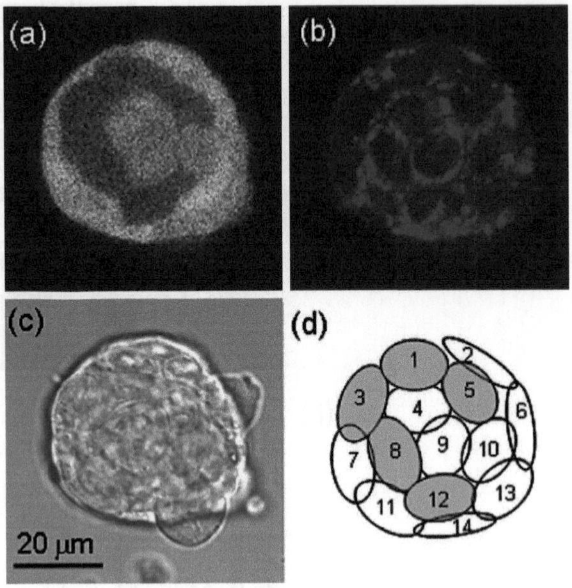

Fig. 13.10. Confocal fluorescent microscopic images of EGFP expression and SYTO 83 staining in G4-2 cells after microinjection of siEGFP. Pressure condition: 0.70 kgf/cm^2, 30 ms. siEGFP$_{\text{C-CAP}}$: 50 ng/μL. (**a**) An EGFP fluorescence image, (**b**) SYTO 83-stained image, (**c**) a phase-contrast image, (**d**) distribution of knocked-down cells (*gray colored*). If the injected siEGFP is distributed homogeneously in the injected single-cell, siEGFP should be partitioned symmetrically during the following cell divisions. Consequently all cells of the colony should be knocked-down homogeneously. If the first cell division occurs before homogeneous distribution of siEGFP, the concentrations of siEGFP in the daughter cells will differ because of asymmetric partition. If one daughter cell contains no siEGFP, then all cells generated by subsequent cell divisions will not contain siEGFP. Consequently half of the colony will comprise siEGFP-free cells and the distribution of knocked-down cells will be heterogeneous. The colony shown in **Fig. 13.7** appears to follow this pattern. Notably, knocked-down cells are connected in 3-dimensions, suggesting some order and sites of cell divisions. (Reproduced from Ref. (8) with permission from Springer)

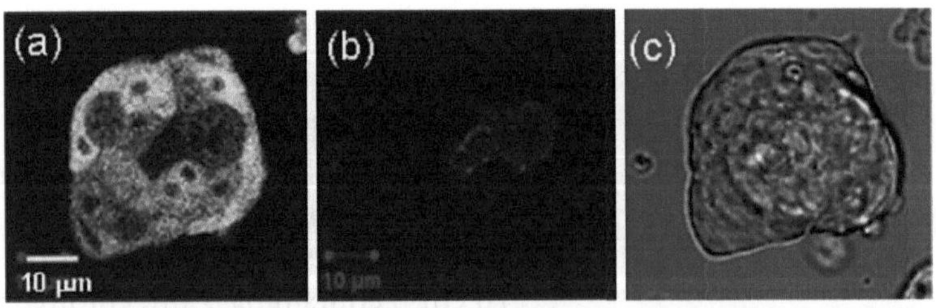

Fig. 13.11. Confocal fluorescent microscopic image of SHP-EGFP expression in SHP-G cells after microinjection of p-shSHP into a target cell. The image was taken 48 h after microinjection of 50 ng/μL p-shSHP and 50 ng/μL p-CAG-IRES-DsRed. (**a**) SHP-EGFP expression image, (**b**) DsRed expression image, (**c**) phase contrast image.

4. Isolate total RNA from the liver of a CD1 female mouse (aged 40 days). Prepare total cDNA according to a standard protocol. Design primers for the PCR to amplify the SHP open reading frame using the full-length cDNA mouse database FANTOM, RIKEN (13). Use primer sets (i) and (ii) for the first PCR and the nested PCR, respectively.

   Primer set (i) for SHP: 4–842 (839 bp)

   Forward 5′–GAGGGCCAGATAGCTGGGAA–3′ (20 mer)

   Reverse 5′–CTCCATTCCACGGGTCACCT–3′ (20 mer)

   Primer set (ii) for SHP: 16–825 (810 bp)

   Forward 5′–GCTGGGAAGAAACAGGAACA–3′ (20 mer)

   Reverse 5′–CCTCAGCAAAAGCATGTCTTC–3′ (21 mer)

5. Prepare all solutions with sterilized pure water that has a conductivity no greater than 5 μmho. In the case of washing of glass capillaries, pure water does not need to be sterilized.
6. Perform lipofection by adding a mixture of lipofectamine 2000 (Invitrogen Japan K.K.) and siRNA to a 24-well multidish. Adjust the lipofectamine concentration at 2% v/v and vary the concentration of siRNA from 1to 100 nM.
7. SMSR is constantly upgraded. Contact H. Matsuoka (mhide@cc.tuat.ac.jp) about the most modern model of SMSR and its commercial availability. SMSR may be replaced by other apparatus with a similar function. Microinjection devices are available from several companies, as are automatic stages for microscopes. Repetitive microinjections may be performed by an expert with such apparatus. However, in our experience, the following points are important to obtain a success rate as high as 10% for ES cells:

   (1) Determine the optimal conditions for capillary tip shape and pressure. The ejected quantity should be sensitive to these conditions.

   (2) The experimenter should be able to concentrate solely on injection of the cells. Preregistration of target cells and the automated calling of each cell to the view center by a one-click operation enable this operational approach.

   (3) The XY address of each target cell must be registered with 2–3 μm precision. When a culture dish is removed from the automatic stage and later returned, it is important to be able to locate the target cells at their originally registered positions as precisely as possible.

Acknowledgments

We thank H. Niwa of the Center for Developmental Biology, RIKEN, Kobe, Japan for providing us with feeder free mouse ES cells (EB3 and G4-2) and plasmids (pCAG-IRES-DsRed and pCAG-cHA-IP-EGFP). This work was supported by funding to H. Matsuoka from CREST of Japan Science and Technology Agency on the research subject "The High Throughput Creation of Disease Model Cells and the Analysis of Their Function." This work was partially supported by Strategic Research Promotion Program, the Ministry of Education, Culture, Sports, Science, and Technology, on the research subject "Development of Next Generation Bioresources."

References

1. Matsuda, T., Nakamura, T., Nakao, K., Arai, T., Katsuki, M., Heike, T., et al. (1999) STAT3 activation is sufficient to maintain an undifferentiated state of mouse embryonic stem cells. *EMBO J* **18**, 4261–4269.
2. Niwa, H., Miyazaki, J., and Smith, A. G. (2000) Quantitative expression of Oct-3/4 defines differentiation, dedifferentiation, or self-renewal of ES cells. *Nat Genet* **24**, 372–376.
3. Niwa, H. (2007) How is pluripotency determined and maintained? *Development* **134**, 635–646.
4. Avilion, A. A., Nicolis, S. K., Pevny, L. H., Perez, L., Vivian, N., and Lovell-Badge, R. (2003) Multipotent cell lineages in early mouse development depend on Sox-2 function. *Genes Dev* **17**, 126–140.
5. Pan, G., and Thomson, J. A. (2007) Nanog and transcriptional networks in embryonic stem cell pluripotency. *Cell Res* **17**, 42–49.
6. Matsuoka, H., Komazaki, T., Mukai, Y., Shibusawa, M., Uetake, N., Saito, M., et al. (2005) High throughput easy microinjection with a single-cell manipulation supporting robot. *J Biotechnol* **116**, 185–194.
7. Matsuoka, H., and Saito, M. (2006) High throughput microinjection technology toward single-cell bioelectrochemistry. *Electrochemistry* **74**, 12–18.
8. Matsuoka, H., Shimoda, S., Ozaki, M., Mizukami, H., Shibusawa, M., Yamada, Y., et al. (2007) Semi-quantitative expression and knockdown of a target gene in single-cell mouse embryonic stem cells by high performance microinjection. *Biotechnol Lett* **29**, 341–350.
9. Nishigori, H., Tomura, H., Tonooka, N., Kanamori, M., Yamada, S., Sho, K., et al. (2001) Mutations in the small heterodimer partner gene are associated with mild obesity in Japanese subjects. *Proc Natl Acad Sci USA* **98**, 2575–2580.
10. Bavner, A., Sanyal, S., Gustafsson, J. K., and Treuter, E. (2005) Transcriptional corepression by SHP: Molecular mechanisms and physiological consequences. *Trends Endocrinol Metab* **16**, 478–488.
11. Yamada, Y., Yamaguchi, N., Ozaki, M., Shinozaki, Y., Saito, M., and Matsuoka, H. (2008) Instant cell recognition system using microfabricated coordinate standard chip useful for combinable cell observation with multiple microscopic apparatus. *Microsc Microanal* **14**, 236–242.
12. Ui-Tei, K., Naito, Y., Takahashi, F., Haraguchi, T., Ohki-Hamazaki, H., Juni, A., et al. (2004) Guidelines for the selection of highly effective siRNA sequences for mammalian and chick RNA interference. *Nucleic Acids Res* **32**, 936–948.
13. RIKEN FANTOM Clone ID: 1300007F22 (Nr0b2: nuclear receptor subfamily 0, group B, member 2, full insert sequence) (http://fantom3.gsc.riken.jp/db/annotate/main.cgi?masterid=1300007F22)

Section III

MicroRNAs

Chapter 14

Preparation and Analysis of MicroRNA Libraries Using the Illumina Massively Parallel Sequencing Technology

Ryan D. Morin, Yongjun Zhao, Anna-Liisa Prabhu, Noreen Dhalla, Helen McDonald, Pawan Pandoh, Angela Tam, Thomas Zeng, Martin Hirst, and Marco Marra

Abstract

MicroRNAs are key regulators of gene expression in diverse biological processes and their importance in embryonic stem cells is indisputable. New 'next-generation' technologies such as Illumina massively parallel sequencing offer vast improvements, in both scale and sensitivity, to microRNA profiling studies. We describe a detailed procedure for the preparation of small RNA libraries for Illumina sequencing. We further comment on approaches for analyzing the resultant sequence data for measuring microRNA abundance.

Key words: Embryonic stem cell, adapter, solexa, ligation, sequence alignment.

1. Introduction

MicroRNAs (miRNAs) are short (19–25 nucleotides) RNA molecules that derive from stable fold-back substructures of larger transcripts. Primary miRNA transcripts (pri-miRNAs) are cleaved by a complex of Drosha and its cofactor DGCR-8 producing one or more miRNA precursor (pre-miRNA) hairpins with a 2-nt overhang at their 3′ ends. Pre-miRNAs are exported from the nucleus to the cytoplasm by Exportin-5, after which they are further cleaved by Dicer, releasing short RNA duplexes. After cleavage by Dicer, one of the strands of the remaining RNA duplex is thought to be preferentially incorporated into the RNA-induced silencing complex (RISC), leaving an inactive

B. Zhang, E.J. Stellwag (eds.), *RNAi and MicroRNA-Mediated Gene Regulation in Stem Cells*, Methods in Molecular Biology 650, DOI 10.1007/978-1-60761-769-3_14, © Springer Science+Business Media, LLC 2010

molecule (miRNA*) that is rapidly degraded (1). Hybridization of miRNAs (within RISC) to their target mRNAs can result in posttranscriptional regulation by two separate mechanisms: transcript degradation or translational repression. MiRNA-mediated regulation is involved in key processes such as differentiation and the cell cycle and, as such, numerous miRNAs have been implicated in the maintenance of embryonic stem cells (2). Considering that 20 new human miRNAs have been added to miRBase in the past few months alone (based on miRBase 13)(3), it is clear that we still lack a complete list of the human miRNA genes. Additionally, evaluation of expression levels of the known miRNAs amongst the numerous tissue and cell types (both normal and diseased) is a field still in its infancy. This field has recently been accelerated by the commercialization of massively parallel sequencing (2, 4). With the depth of sequencing now possible, we have an opportunity to identify low-abundance miRNAs or those exhibiting modest expression differences between samples.

Massively parallel sequencing strategies allow the simultaneous 'reading' of sequences of up to millions of cDNAs derived from small RNA fragments. The Illumina (previously known as Solexa) sequencing approach is capable of producing read numbers that are orders of magnitude greater (5) than that of 454 (Roche) sequencing. This technology involves universal oligonucleotides (oligos) that are physically anchored to a fixed surface inside lanes of a 'flowcell.' A small RNA library is first prepared by enzymatically ligating 5′ and 3′ adapters to small RNA molecules, and enriched by 10–15 cycles of PCR using primers that contain sequences homologous to these adapters and complementary to the oligos on the flow cell surface. In the final library construct, the 5′ end contains a sequencing primer binding sequence. A detailed procedure of the library preparation process is described in this chapter. Once completed, a small RNA library can be loaded into one or more flowcell lanes and, by bridge amplification, produce clusters comprising double-stranded DNA each derived from a single template molecule. Each cluster is individually sequenced in parallel by successive addition of the four nucleoside triphosphates. These nucleotides have a 3′ reversible terminator moiety that is also a fluorophore. This prevents successive incorporation of multiple nucleotides in polynucleotide tracts, a major issue with early 454 (Roche) sequencing. Images of the laser-excited fluorophores are captured after each nucleotide addition step and then the fluorophore is chemically cleaved to allow further extension reactions. An overlay of all images collected during this process is used to produce full-length sequence reads. Because the diversity of the microRNAome is much less than the number of reads, many identical sequences are produced (and can be considered multiple observations). By considering each read as a single observation of a molecule of that microRNA, massively parallel sequence data provide the expression profile of a sample while offering the potential to reveal sequence discrepancies between microRNAs and the human reference genome. However, with only a few exceptions (2, 6), microRNA reads that do not match the genome perfectly are generally discarded due to difficulties in alignment and separating the real observations from noise arising from difficulties in alignment to the human reference genome or sequencing errors. This mindset was encouraged by the knowledge that each individual sequence read had a high error rate (relative to classical capillary sequence data). Next-generation alignment softwares, such as BWA or Novoalign, are capable of rapidly finding optimal alignments in the human genome while allowing multiple mismatches and variable read length (7). Considering that the variability of microRNAs at the sequence level is poorly understood, routine global analysis of reads (including those with imperfect alignments) is important for proper quantitation of microRNAs and discovery of known or novel sequence variants. The latter portion of this chapter describes our computational approach to analyze Illumina sequencing libraries.

1.1. MicroRNA Library Preparation

This version of the library construction process is divided into 7 days to fit work schedules, some days such as gel excision requires only a short time investment. To prevent contamination, the process is segregated into three areas, pre-PCR including RNA and cDNA synthesis, library construction, and post-PCR. Typically, 4–8 samples can be processed in parallel allowing higher productivity; however, attention must be paid to prevent sample swap and potential cross contamination. New versions of these protocols with substantial reduction in manual labor are currently under development.

2. Materials

Materials used in this protocol are listed by day. Comparable supplies may be purchased from other sources at the users' choice.

2.1. Library Preparation, Day 1

1. RNAse Zap (Ambion) (used on subsequent days, 2–7).
2. Wet ice (used on subsequent days, 2–7).
3. RNAse-free 2-mL microcentrifuge tubes (used on subsequent days 2–7).
4. RNAse-free 0.5-mL nonstick microcentrifuge tubes (used on subsequent days, 2–7).
5. Micropipettors: P2, P10, P20, P200, and P1000 (used on subsequent days, 2–7).
6. Barrier micropipettor tips, sterile: 10, 20, 200 and 1000 μL (used on subsequent days, 2–7).

7. Mini-centrifuge (used on subsequent days, 2–7).
8. Vortex mixer (used on subsequent days, 2–7).
9. DEPC water (used on subsequent days, 2–7).
10. SybrGreen II (Cambex Bio).
11. Novex TBE electrophoresis buffer, 5X.
12. Novex TBE-Urea sample buffer.
13. XCell *SureLock* Mini-Cell.
14. 10% TBE-Urea Gels, 1.0 mm, 10 wells.
15. 14–30 ssRNA Ladder Marker.
16. 10 bp DNA Ladder.
17. 18 gauge needle.
18. 5 M NaCl.

2.2. Library Preparation, Day 2

1. Thermomixer 1.5 mL (Eppendorf) (used on subsequent days, 3–7).
2. Spin-X Filter columns.
3. Mussel glycogen.
4. Ethanol (100% EtOH).
5. 5′ RNA adapter (Illumina).
6. Ethanol (70% EtOH).
7. T4 RNA ligase 5 U/μL.
8. RNase Out (Invitrogen).

2.3. Library Preparation, Day 3

1. SybrGreen II (Cambex Bio).
2. Novex TBE Electrophoresis Buffer 5X.
3. XCell *SureLock* Mini-Cell (Invitrogen).
4. 10% TBE-Urea gels, 1.0 mm, 10-well.
5. 5 M NaCl.
6. 18-gauge needle.

2.4. Library Preparation, Day 4

1. Spin-X Filter Columns.
2. Mussel glycogen.
3. Ethanol (100% EtOH).
4. Ethanol (70% EtOH).
5. 3′ RNA adapter (Illumina).
6. T4 RNA ligase, 5 U/μL.
7. RNase Out (Invitrogen).

2.5. Library Preparation, Day 5

1. SybrGreen II (Cambex Bio).
2. Novex TBE electrophoresis buffer, 5X.
3. Novex TBE-Urea sample buffer, 2X.

4. 10% TBE-Urea gels, 1.0 mm, 10-well.
5. XCell *SureLock* Mini-cell (Invitrogen).
6. 10 bp DNA ladder.
7. 5 M NaCl.
8. 18-gauge needle.

2.6. Library Preparation, Day 6

1. dNTP Mix, 10 mM each.
2. 0.1 M DTT.
3. RNase Out, 40 U/μL (Invitrogen).
4. Superscript II Reverse Transcriptase (Invitrogen).
5. Mussel glycogen.
6. Ethanol (100% EtOH).
7. 18 Mega Ohm water.
8. Ethanol (70% EtOH).
9. Small RNA RT Primer, 100 μM (Illumina).
10. 5x Phusion HF buffer.
11. Small RNA PCR primer 1 (Illumina).
12. Small RNA PCR primer 2 (Illumina).
13. Phusion Hot Start High-fidelity DNA Polymerase (New England Biolabs).
14. 8% TBE Gels, 1.0 mm, 10 well.
15. 1x TBE buffer.
16. XCell *SureLock* Mini-Cell (Invitrogen).
17. 10x BPB/XC loading buffer.
18. Cling wrap.
19. 25 bp DNA ladder.
20. SybrGreen I (CAMBEX Bio Science).
21. Peltier Thermal Cycler.
22. Power Supply, LVC2kW, 48volts DC.
23. Razor blades, single-edged.
24. Power supply.
25. Typhoon 9400 scanner (Amersham).
26. Dark Reader/transilluminator.

2.7. Library preparation, Day 7

1. Heat block.
2. Nutator shaker – Clay Adams brand (VWR).
3. Spin-X Filter Columns.
4. DNAAWAY (MBS).
5. Mussel glycogen, 20 mg.

6. 7.5 M Ammonium acetate.
7. Ethanol (100% EtOH).
8. 70% EtOH.
9. EB buffer, PCR Purification kit (Qiagen).

3. Methods

As the standard Illumina miRNA library starts with gel excision of 14–30 nt RNA from total RNA serving as small RNA enrichment, a total RNA extraction method such as TRIZOL usually works perfectly. Column-based enrichment methods are not necessary in this protocol. Dedicated work stations, solutions, and equipment free from nucleases are required. Any contamination, especially RNases, may lead to library failure. It is recommended that input RNA be checked by Agilent Bioanalyzer RNA Nanochip to ensure RNA Integrity Number (RIN) is greater than 7.

3.1. Library Preparation, Day 1

Purpose: To fractionate 10 μg of sample total RNA on a 10% TBE-Urea gel and excise the small RNAs in the size range of 14–30 nucleotides.

3.1.1. Preparing and Electrophoresing a Precast 10% TBE-Urea Gel at MicroRNA Gel Workstation

1. Put on a clean pair of gloves and lab coat.
2. Wipe down the workbench, small equipment, and ice bucket with RNAse Zap.
3. Wipe down with DEPC-water.
4. Lay down new bench coat.
5. Preheat a heat block to 70°C.
6. Dilute stock 5X Novex TBE buffer using DEPC water to 1X electrophoresis buffer. The gel apparatus will require approximately 400 mL buffer, and an additional 100 mL is required for gel staining.
7. Cut open the Novex gel cassette pouch to remove the gel cassette, and drain away the gel packaging buffer. Handle the gel cassette by the edges only. Rinse the gel cassette with RNase-free deionized water.
8. Peel off the tape covering the slot on the back of the gel cassette.
9. In one fluid motion, pull the sample well former out of the cassette.
10. Use a 1-mL pipette to gently wash the cassette wells with 1X electrophoresis buffer. Repeat twice, and then fill the sample wells with electrophoresis buffer.

11. Assemble the gel apparatus as follows:
    (A) Lower the buffer core into the lower buffer chamber so that the negative electrode fits into the opening in the gold plate on the lower buffer chamber.
    (B) Insert the gel tension wedge into the XCell *SureLock* cell behind the buffer core. Make sure the gel tension wedge is in its unlocked position, allowing the wedge to slip easily into the XCell *SureLock* unit.
    (C) Insert the gel cassette into the lower buffer chamber in front of the core, with the well side of the cassette facing in towards the buffer core. The slot on the back of the cassette must face out towards the lower buffer chamber. Place the buffer dam behind the core.
    (D) Pull forward on the gel tension lever in a direction towards the front of the XCell *SureLock* unit until lever comes to a firm stop and the gel/buffer dam appear snug against the buffer core.
12. Fill the upper buffer chamber with 200 mL of the 1X TBE electrophoresis buffer. Ensure that the upper buffer chamber is not leaking. If the level of the electrophoresis buffer drops, the apparatus will need to be reassembled.
13. Fill the lower buffer chamber with approximately 200 mL of electrophoresis buffer through the gap between the gel tension wedge and the back of the lower buffer chamber as shown below:
14. Align the lid on the buffer core. The lid can only be firmly seated if the (–) electrode is aligned over the banana plug on the right. Caution: Power must be off before connecting the XCell *SureLock* Mini Cell to the power supply.
15. Turn on the power. Pre-electrophorese the gel at 180 V for 15–20 min.
16. Prepare the sample by diluting 10 μg of total RNA with an equal volume of Novex TBE-Urea Sample Buffer (2X).
17. For each gel to be electrophoresed, add an equal volume (5 μL) of Novex TBE-Urea Sample Buffer (2X) to 5 μL aliquots of diluted DNA and RNA ladders. The DNA ladder is optional as it is less accurate owing to its faster gel migration relative to the RNA ladder.
18. Immediately prior to loading, heat the sample and ladders at 70°C for 3 min, and then chill on ice.
19. After the gel has pre-electrophoresed, stop the power and remove the lid.
20. Carefully rinse out the wells with 1X TBE electrophoresis buffer using a 1-mL pipette tip.

21. Using a P10 tip, load the DNA and RNA ladders into sample wells on the left side of the gel. Leave a gap of approximately 5 wells, and carefully load the sample into as many wells as required, keeping in mind that one well holds a maximum volume of 25 μL.
22. Replace the lid of the apparatus, and start the gel electrophoresis.
23. Electrophorese the 10% TBE-Urea gel at 180 V for 45 min. The bromophenol blue dye corresponds to the 20 nucleotide DNA marker, and it will have migrated approximately ³⁄₄ of the length of the gel.

3.1.2. Staining the TBE-Urea Gel and Excision of the Small RNA Fraction

1. Prepare sets of tubes for shearing the gel slices containing the small RNA fraction. Approximately one set of tubes is needed per lane loaded. Make a hole through the bottom of a 0.5-mL RNase-free nonstick tube with an 18-gauge needle and place on top of a 2-mL RNase-free tube.
2. Prepare fresh 1X TBE/SybrGreenII stain; 10 μL stock in 100 mL 1x TBE in a clean tray designated for staining RNA gels. Minimize exposure to light.
3. At the end of the gel electrophoresis, turn off the power and disconnect the cables from the power supply.
4. Remove the lid and unlock the gel tension lever. The gel tension wedge can be left in place.
5. Remove the gel cassette from the mini-cell. Handle gel cassettes by their edges only.
6. Lay the gel cassette (well side up) on a flat surface, such as the benchtop. Carefully insert the gel knife's beveled edge into the narrow gap between the two plates of the cassette.
7. Push up and down gently on the knife's handle to separate the plates. You will hear a cracking sound, which means you have broken the bonds that hold the plates together. Repeat until you have broken the bonds on one side. Rotate the cassette and repeat, until the two plates have completely separated.
8. Upon opening the cassette, the gel may adhere to either side. Remove and discard the plate without the gel, allowing the gel to remain on the other plate.
9. Remove the gel from the cassette plate by one of two methods:
   A. If the gel remains on the shorter (notched plate), use the sharp edge of the gel knife to remove the bottom foot of the gel. Hold the gel knife at a 90° angle to the gel and the slotted cassette plate. Push straight down on the knife to cut the gel. Repeat the motion across the

gel to cut the entire foot. Hold the cassette plate and gel over the gel-staining container with the gel facing downward. Use the knife to carefully loosen one lower corner of the gel and allow the gel to peel away from the plate.

B. If the gel remains on the longer (slotted) plate, hold the cassette plate and gel over the gel-staining container with the gel facing downward. Gently push the gel knife through the slot in the cassette, until the gel peels away from the plate.

10. Stain the gel for 10 min with gentle agitation.
11. Wipe down the gel imager surface (Typhoon scanner or other gel document system) with DEPC water and ensure that it is completely clear of all gel debris.
12. Scan the gel and save the image.
13. Carefully remove the gel and lay onto a clean piece of cling wrap on a dark reader.
14. Using a brand new razor blade, excise the gel fraction of the sample corresponding to the 14–30 nucleotide region (*see* **Fig. 14.1**). Use the RNA marker as a reference for cutting. Excise a gel slice that corresponds to the region between the lowest and highest RNA marker band. The RNA ladder comprises 5 fragments; however, two of the fragments do not resolve in the 10% precast TBE gel and a pattern of 4 fragments is observed.
15. Transfer gel slices into the prepared 0.5 mL tubes for gel shearing, using approximately one tube per lane loaded.
16. Close the 0.5-mL tube lids to contain the gel slices, but leave the 2-mL tube lids open to centrifuge the sample. Centrifuge at 12,000 rpm at 4°C for 3 min. The gel slice should shear through the hole and collect into the bottom of the 2-mL tubes.
17. Discard the 0.5-mL tubes and add 400 μL of 0.3 M NaCl to each 2-mL tube. Vortex gently.
18. Transfer the tubes to a 4°C RNA refrigerator for overnight incubation.
19. Clean the XCell *SureLock* apparatus with tap water and Micro90, and rinse thoroughly with tap water. Then rinse thoroughly with DEPC water.
20. Clean the RNA gel electrophoresis area.

3.2. Library Preparation, Day 2

Purpose: To purify and precipitate the small RNA fraction excised from the denaturing PAGE gel electrophoresed on Day 1 of Micro RNA library construction and to set up the 5′ adapter ligation reaction.

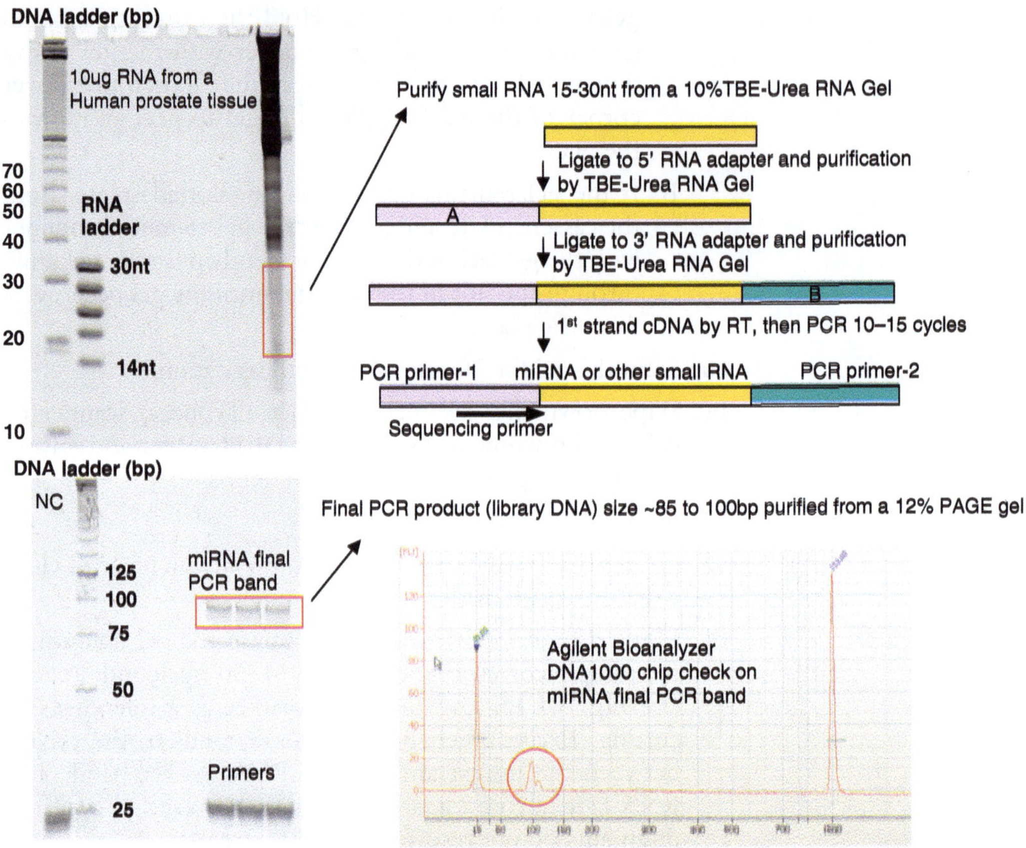

Fig. 14.1. MicroRNA Library Construction Work Flow. 5–10 μg of total RNA is size fractionated on a 10% TBE-Urea gel, and a 15–30 nt fraction is excised. RNA is extracted from the gel slice and precipitated. The pellet is washed and air dried, and resuspended in DEPC water. A 5′ RNA adapter is ligated to the RNA using T4 RNA ligase overnight. The ligated RNA is size fractionated on a 10% TBE-Urea gel and a 40–65 nt fraction is excised, eluted from the gel, and once again precipitated. The 3′ RNA adapter is subsequently ligated to the precipitated RNA. Ligated RNA is size fractionated on a 10% TBE-Urea gel and the 60–95 nt fraction excised. After elution and precipitation, the RNA is converted to single-stranded cDNA using Superscript II reverse transcriptase (Invitrogen) and Illumina's small RNA RT-Primer following the manufacturer's instructions. The resulting cDNA is PCR-amplified with Hotstart Phusion DNA Polymerase (BioRad) in 10–15 cycles using Illumina's small RNA primer set. PCR products are gel purified, precipitated, resuspended, and quantified using an Agilent DNA 1000 chip (Agilent). The quantified PCR products are diluted to 15 nM stock, which and is further diluted to the appropriate working concentration, denatured for cluster generation to the desired density, and sequenced on the Illumina Genome Analyzer.

3.2.1. Small RNA Purification and Precipitation

1. Put on a clean pair of gloves and lab coat.
2. Wipe down the workbench, small equipment, and ice bucket with RNase Zap.
3. Lay down new bench coat.
4. Retrieve fresh ice and all required reagents.
5. Retrieve the sample tubes containing the gel slurry of the small RNA fraction isolated from a denaturing PAGE gel on the previous day.

6. Vortex the tubes, pulse-centrifuge, and transfer the gel slurry from each tube onto the top of a Spin-X filter column.
7. Centrifuge the sample through the spin column into the collection tube at 12,000 rpm for 4 min at 4°C.
8. Check each spin column tube and ensure all the buffer has spun through the filter. Recentrifuge the tubes if there is still liquid trapped in the gel material.
9. Transfer the eluate from one sample into a single nonstick RNase-free 1.5-mL tube, up to a maximum total volume of 400 μL. The total volume of eluate will depend on the amount of excised gel and volume of added elution buffer. Adjust the amount of ethanol used to precipitate the small RNAs accordingly, and add the reagents to the eluate as follows:
   - 400 μL Eluate
   - 3 μL Mussel glycogen
   - Ethanol (100% EtOH)
10. Vortex, then incubate at −80°C for a minimum of 20–30 min.
11. Centrifuge at 14,000 rpm for 30 min at 4°C in a microcentrifuge.
12. Discard the supernatant by carefully decanting the EtOH supernatant into a new microcentrifuge tube, being careful to prevent the pellet from sliding out of the tube.
13. Wash the pellet with 800 μL of 70% EtOH by adding the EtOH solution and inverting the tube. Centrifuge at 14,000 rpm at 4°C for 2 min to repellet. Discard the supernatant as previously.
14. Repeat the 70% EtOH wash.
15. Pulse-centrifuge the sample tube and carefully remove any residual ethanol by using a P200 pipette tip first to remove the majority of the supernatant, then finally using a P10 pipette tip to remove the last trace of solution. Mark the outside bottom of the tube to better locate the pellet when resuspending in DEPC water.
16. Allow the tube to air-dry for approximately 5–10 min at room temperature, until the white precipitate is no longer visible.
17. Resuspend the pellet in 6.2 μL of DEPC water and place on ice.

3.2.2. 5′ Adapter Ligation Reaction

1. Set up the 5′ Adapter ligation as per **Table 14.1**.
2. Mix by pipetting up and down slowly. Pulse-centrifuge.

Table 14.1
5′ Adapter ligation reaction mixture

Reagent	Amount (μL)
Purified small RNA fraction	6.2
5′ Illumina RNA Adapter, 5 μM	1.3
10X Ligation Buffer	1
T4 RNA Ligase (Ambion, 5 U/μL)	1
RNase Out (Invitrogen, 40 U/μL)	0.5
Total reaction volume (μL)	10

3. Incubate at 16°C overnight.
4. Remove, rinse in dH_2O, and dry the previous day's decontamination soak items.
5. Place used racks in fresh bleach decontamination soak.
6. Clean the RNA workstation area. Wipe down the workstation, pipettors, and other small equipment with RNaseZap.

3.3. Library Preparation, Day 3

Purpose: To separate the 5′ adapter-ligated RNA products from excess adapters and unligated RNAs on a 10% TBE-Urea gel and to excise the desired sized ligation products (40–65 nt) from the gel.

3.3.1. Preparing and Electrophoresing a Precast 10% TBE-Urea Gel

1. Put on a clean pair of gloves and disposable lab coat.
2. Use RNase ZAP to wipe down the RNA workstation, pipettors. and small equipment.
3. Wipe down with DEPC-water.
4. Lay down new benchcoat.
5. Preheat a heat block to 70°C.
6. Set up one Novex 10% TBE-Urea gel in the XCell *SureLock* Mini-Cell Apparatus according to the instructions outlined in the Illumina microRNA Day 1 library construction protocol.
7. Turn on the power. Pre-electrophorese the gel at 180 Volts for 15–20 min.
8. Retrieve the sample from the overnight incubation reaction at 16°C.
9. Prepare the sample for loading by dilution with an equal volume of Novex TBE-Urea Sample Buffer (2X).
10. Add an equal volume of Novex TBE-Urea Sample Buffer to an aliquot of diluted 10-bp DNA ladder from Invitrogen (5 μL of a 1:25 [40 ng/μL] stock dilution).

11. Prior to loading, heat the sample and DNA ladder at 70°C for 3 min, and then chill on ice.
12. After the gel has pre-electrophoresed, stop the power and remove the lid.
13. Carefully rinse out the sample wells with 1X electrophoresis buffer using a 1-mL pipette tip.
14. Using a P10 tip, load the DNA ladder into a well on the left side of the gel. Leave a gap of approximately 5 wells, and carefully load the sample into a single well on the right side of the gel, keeping in mind that one well holds a maximum volume of 25 μL.
15. Replace the lid of the apparatus, and start the gel electrophoresis.
16. Electrophorese the 10% TBE-Urea gel at 180 V for 55 min, until the bromophenol blue has migrated to almost the bottom of the gel.

3.3.2. Staining the TBE-Urea Gel and Excision of the 5′ Adapter-Ligated RNA Products

1. Prepare tubes for shearing the gel slice containing the Adapter Ligation products. Make a hole through the bottom of a 0.5-mL RNase-free nonstick tube with an 18-gauge needle and place on top of a 2-mL RNase-free tube.
2. Prepare fresh 1X TBE/SybrGreenII stain; 10 μL stock in 100 mL 1x TBE in a clean tray designated for staining RNA gels. Minimize exposure to light.
3. At the end of the gel electrophoresis, turn off the power and disconnect the cables from the power supply.
4. Remove the lid and unlock the gel tension lever. The gel tension wedge can be left in place.
5. Remove the gel cassette from the mini-cell. Handle gel cassettes by their edges only.
6. Lay the gel cassette (well side up) on a flat surface, such as the benchtop. Carefully insert the gel knife's beveled edge into the narrow gap between the two plates of the cassette.
7. Push up and down gently on the knife's handle to separate the plates. You will hear a cracking sound which means you have broken the bonds which hold the plates together. Repeat until you have broken the bonds on one side. Rotate the cassette and repeat, until the two plates are completely separated.
8. Upon opening the cassette, the gel may adhere to either side. Remove and discard the plate without the gel, allowing the gel to remain on the other plate.
9. Remove the gel from the cassette plate using the same method described in Day 1.

10. Stain the gel using 1X TBE/SybrGreenII similarly prepared as in Day 1 for 10 min with gentle agitation.
11. Wipe down the imager (scanner) surface with DEPC Water and remove any pieces of gel material.
12. Scan the gel and save the image.
13. Transfer the gel to a fresh piece of cling wrap on the dark reader. Using a brand new razor blade excise the region of the ligation reaction products that corresponds to 40–65 nucleotides of the DNA ladder.
14. Transfer the gel slice into the 0.5-mL tube.
15. Using another tube as a balance and with the lids tailing (left of tube position), centrifuge at 12,000 rpm at 4°C for 3 min. The gel slice should shear through the hole and collect into the bottom of the 2-mL tube.
16. Discard the 0.5-mL tube and add 400 μL of 0.3 M NaCl to the gel slurry in the 2-mL tube. Vortex gently.
17. Transfer the tube to a 4°C refrigerator for overnight incubation.

3.4. Library Preparation, Day 4

Purpose: To purify and precipitate the 5′ adapter-ligated small RNA products excised from the denaturing PAGE gel electrophoresis on Day 3 and to set up the 3′ adapter ligation reaction.

3.4.1. Purification and Precipitation the 5′ Adapter-Ligated Small RNA Products

1. Put on a clean pair of gloves and lab coat.
2. Wipe down the workbench, small equipment, and ice bucket with RNAse Zap.
3. Lay down new bench coat.
4. Retrieve fresh ice and all required reagents.
5. Retrieve the sample tubes containing the gel slurry of the 5′ adapter-ligated small RNA products isolated from a denaturing PAGE gel on Day 3 of Illumina Micro RNA library construction. Vortex the tubes, pulse-centrifuge, and transfer the gel slurry from each tube onto the top of a Spin-X filter, and centrifuge at 12,000 rpm for 4 min at 4°C.
6. Check each spin column tube and ensure that all of the buffer has spun through the filter. Recentrifuge the tubes if there is still liquid trapped in the gel material.
7. Transfer all eluate for one sample into one nonstick RNase-free 1.5-mL tube. The total volume of eluate will depend on the amount of excised gel and volume of added elution buffer; adjust the following amounts of reagents used to precipitate the ligation products accordingly, and add the reagents to the eluate as follows:

- 400 μL Eluate
- 3 μL Mussel glycogen
- Ethanol (100% EtOH)

8. Vortex, then incubate at −80°C for a minimum of 20–30 min.
9. Centrifuge at 14,000 rpm for 30 min at 4°C in a microcentrifuge.
10. Discard the supernatant by carefully decanting the EtOH supernatant into a new microcentrifuge tube, being careful to prevent the pellet from sliding out of the tube.
11. Wash the pellet with 800 μL of 70% EtOH by adding the EtOH solution and inverting the tube. Centrifuge at 14,000 rpm at 4°C for 2 min. Discard the supernatant as previously.
12. Repeat the 70% EtOH wash.
13. Pulse-centrifuge the sample tube and carefully remove any residual ethanol by using a P200 pipette tip first to remove the majority of the supernatant, then finally using a P10 pipette tip to remove the last trace of solution. Mark the outside bottom of the tube to better locate the pellet when re-suspending in DEPC water.
14. Allow tube to air-dry for approximately 5–10 min at room temperature, until the white precipitate is no longer visible.
15. Resuspend the pellet in a total volume of 6.3 μL DEPC water per sample.

3.4.2. 3′ Adapter Ligation Reaction

1. Set up the 3′ adapter ligation reaction as per **Table 14.2**.
2. Mix by pipetting up and down slowly. Pulse-centrifuge. Incubate at 16°C overnight.
3. Wipe down the workstation, pipettors, and small equipment with RNase Zap.
4. Remove, rinse in dH_2O, and dry the previous day's decontamination soak items.
5. Place used racks in a fresh bleach decontamination soak.

3.5. Library Preparation, Day 5

Purpose: To separate the 5′, 3′ adapter-ligated small RNA products from excess adapters on a 10% TBE-Urea gel, and to excise the desired sized ligation products (60–95 nt) from the gel.

3.5.1. Isolation of the 5′, 3′ Adapter-Ligated Small RNA Products in a TBE-Urea Gel

1. Put on a clean pair of gloves and disposable lab coat.
2. Use RNAse ZAP to wipe down the RNA workstation, pipettors, and small equipment.

Table 14.2
3′ Adapter ligation reaction mixture

Reagent	Amount (μL)
Purified 5′ ligation product	6.3
3′RNA Adapter	1.2
10X Ligation Buffer	1
T4 RNA Ligase (5 U/μL)	1
RNase Out, (40 U/μL)	0.5
Total reaction volume (μL)	10

3. Wipe down with DEPC-water.
4. Lay down new benchcoat.
5. Set up one Novex 10% TBE-Urea gel in the XCell *SureLock* Mini-Cell Apparatus according to the instructions outlined in the Illumina Micro RNA Day 1 library construction protocol.
6. Turn on the power. Pre-electrophorese the gel at 180 V for 15–20 min.
7. Retrieve the sample from the overnight incubation reaction at 16°C.
8. Prepare the sample for loading by dilution of the ligation reaction with an equal volume of Novex TBE-Urea Sample Buffer (2X).
9. Add an equal volume of Novex TBE-Urea Sample Buffer (2X) to an aliquot of diluted 10-bp DNA ladder from Invitrogen (5 μL of a 1:25 [40 ng/μL] stock dilution).
10. Immediately prior to loading, heat the sample and DNA ladder at 70°C for 3 min, and then chill on ice.
11. After the gel has pre-electrophoresed, stop the power, and remove the lid.
12. Carefully rinse out the wells with 1X TBE electrophoresis buffer using a 1-mL pipette tip.
13. Using a P10 tip, load the DNA ladder into a well on the left side of the gel. Leave a gap of approximately 5 wells, and load the sample into a single well, keeping in mind that one well holds a maximum volume of 25 μL.
14. Replace the lid of the apparatus, and start the gel electrophoresis.
15. Electrophorese the 10% TBE-Urea gel at 180 V for 55 min, until the bromophenol blue has migrated to almost the bottom of the gel.

3.5.2. Staining the TBE-Urea Gel and Excision of the 5′ Adapter-Ligated RNA Products

1. Prepare tubes for shearing the gel slice containing the Adapter Ligation products. Make a hole through the bottom of a 0.5-mL RNase-free nonstick tube with an 18-gauge needle and place on top of a 2 mL RNAse-free tube.
2. Prepare fresh 1X TBE/SybrGreen II stain; 10 μL stock in 100 mL 1x TBE in a clean tray designated for staining RNA gels. Minimize exposure to light.
3. At the end of the gel electrophoresis, turn off the power and disconnect the cables from the power supply.
4. Remove the lid and unlock the gel tension lever. The gel tension wedge can be left in place.
5. Remove the gel cassette from the mini-cell. Handle gel cassettes by their edges only.
6. Lay the gel cassette (well side up) on a flat surface, such as the bench-top. Carefully insert the gel knife's beveled edge into the narrow gap between the two plates of the cassette.
7. Push up and down gently on the knife's handle to separate the plates. You will hear a cracking sound, which means you have broken the bonds that hold the plates together. Repeat until you have broken the bonds on one side. Rotate the cassette and repeat, until the two plates are completely separated.
8. Upon opening the cassette, the gel may adhere to either side. Remove and discard the plate without the gel, allowing the gel to remain on the other plate.
9. Remove the gel from the cassette plate using the same method described in Day 1.
10. Stain the gel using 1X TBE/SybrGreen II similarly prepared as in Day 1 for 10 min with gentle agitation.
11. Wipe down the imager (scanner) surface with DEPC Water and remove any pieces of gel material.
12. Scan the gel and save the image.
13. Transfer the gel to a fresh piece of cling wrap on the dark reader. Using a brand new razor blade excise the region of the ligation reaction products that corresponds to 60–95 nucleotides of the DNA ladder.
14. Transfer the gel slice into the 0.5-mL tube.
15. Using another tube as a balance, and with the lids tailing (left of tube position), centrifuge at 12,000 rpm at 4°C for 5 min. The gel slice should shear through the hole and collect into the bottom of the 2-mL tube.

16. Discard the 0.5-mL tube and add 400 μL of 0.3 M NaCl to the gel slurry in the 2-mL tube. Vortex gently.
17. Transfer the tube to a 4°C refrigerator for overnight incubation.

3.6. Library Preparation, Day 6

Purpose: To purify the 5′, 3′ adapter-ligated small RNAs from gel slurry, and to use the RNAs as template for RT-PCR, followed by separation of the PCR products on an 8% PAGE gel.

3.6.1. Purification and Precipitation of the 5′ 3′ Adapter-Ligated Products

1. Put on a clean pair of gloves and lab coat.
2. Wipe down the workbench, small equipment, and ice bucket with RNase Zap.
3. Lay down new bench coat.
4. Retrieve fresh ice and all required reagent.
5. Thaw all reagents, vortex, and pulse-spin.
6. Preheat heat blocks to 70°C and 48°C.
7. Retrieve the sample tube(s) containing the gel slurry of the 5′ 3′ adapter-ligated RNA products isolated from a denaturing PAGE gel on the previous day. Vortex the tubes, pulse-spin, and transfer the gel slurry from each tube onto the top of a Spin-X filter column.
8. Centrifuge at 12,000 rpm for 4 min at 4°C.
9. Check each spin column tube and ensure that all of the buffer has spun through the filter. Recentrifuge the tubes if there is still liquid trapped in the gel material.
10. Transfer all eluate for one sample into one nonstick RNase-free 1.5-mL tube. The total volume of eluate will depend on the amount of excised gel and volume of added elution buffer; adjust the following amounts of reagents used to precipitate the ligation products accordingly, and add the reagents to the eluate as follows:
    - 400 μL Eluate
    - 3 μL Mussel glycogen
    - Ethanol (100% EtOH)
11. Vortex, then incubate at −80°C for 20–30 min.
16. Centrifuge at 14,000 rpm for 30 min at 4°C in a microcentrifuge.
17. Discard the supernatant carefully by decanting the EtOH supernatant into a new microcentrifuge tube, being careful to prevent the pellet from sliding out of the tube.
18. Wash the pellet with 800 μL of 70% EtOH by adding the EtOH solution and inverting the tube. Centrifuge at 14,000 rpm at 4°C for 2 min. Decant the supernatant as previously.

19. Repeat the 70% EtOH wash.
20. Allow the pellet to air dry for approximately 5–10 min at room temperature, until the white precipitate is no longer visible.
21. Resuspend the pellet in 4.5 μL DEPC-treated water and place on ice.

3.6.2. Reverse Transcription of Small RNA Adapter-Ligated Products

1. Add the small RNA RT primer to the resuspended RNA as follows:
   - 4.5 μL of purified ligated RNA
   - 0.5 μL of small RNA RT primer, 100 μM
2. Place the tube at 70°C for 2 min to heat-denature the RNA, then remove the tube and pulse-centrifuge.
3. Place the tube in a rack on the bench-top at room temperature. Immediately add the following reagents:
   - 2.0 μL of 5x First strand buffer
   - 0.65 μL of dNTPs (10 mM)
   - 1 μL of DTT (100 mM)
   - 0.5 μL of RNAseOut (40 U/μL)
4. Heat at 48°C for 3 min, remove the tube, and turn down the temperature setting of the heat block to 44°C.
5. Add 1.0 μL of Superscript II RT (200 U/μL). Mix gently and pulse-centrifuge the tube.
6. Incubate at 44°C for 1 h.
7. After the 1-h incubation, place tube on ice.
8. Add 12 μL of DEPC water to the first-strand cDNA products to make a final total volume of approximately 20 μL.

3.6.3. PCR Amplification of Small RNA First-Strand cDNA Products

1. Put on a clean pair of gloves.
2. Wipe down the PCR set-up hood, pipettors, small equipment, and ice bucket with RNase Zap.
3. Thaw all required reagents, vortex, and pulse-spin.
4. Prepare the PCR master mix for all samples and one control in a 1.5-mL tube on ice as outlined in **Table 14.3**.
5. Label individual 0.2-mL thin-walled PCR tubes with sample or control name. Set up two tubes for each sample, and label for 10- and 15-cycle PCR reactions.
6. Dispense 5 μL of water into the master mix only control PCR reaction tube. Add 45 μL master mix to all tubes, adding to the sample tubes first and the remainder to the control tube. Close the lids on the tubes and store on ice.
7. Transfer 10 μL aliquots from each sample of first-strand cDNA (5 μL per PCR reaction) into new nonstick tubes.

Table 14.3
PCR reaction mixture

Reagent	Per reaction (μL)	2X Reactions (μL)	3X Reactions (μL)	5X reactions (μL)
DEPC water	30.75	61.5	92.25	153.75
5x Phusion HF buffer	10	20	30	50
DMSO	1.5	3.0	4.5	7.5
dNTPs (10 mM each)	1.25	2.5	3.75	6.25
PCR Primer 1, 25 μM	0.5	1.0	1.5	2.5
PCR Primer 2, 25 μM	0.5	1.0	1.5	2.5
Phusion DNA Polymerase (hot start), 2 U/μL	0.5	1.0	1.5	2.5
Total volume	45	90	135	225

Store all the remaining first-strand cDNA products at minus 20°C.

8. Wipe down the Tetrad thermal cycler with DNAAway and then dH_20.
9. Turn on the Tetrad.
10. Carefully aliquot the sample (cDNA) into the 0.2-mL PCR tubes containing PCR master brew mix, for a final total of 50 μL, changing gloves between samples.
11. Put the sample and control PCR tubes in the alpha unit, ensuring the lid is properly in place. Start PCR program:
    98° 30 s

    followed by 10 or 15 cycles of:

    98° 10 s

    60° 30 s

    72° 15 min

    followed by 72° 5 min
12. While the PCR reaction is proceeding, set up the gel electrophoresis apparatus described in the next section.

3.6.4. Purification of PCR Product

1. Cut open an 8% TBE Novex gel cassette pouch to remove the gel cassette, and drain away the gel-packaging buffer. Handle the gel cassette by the edges only. Rinse the gel cassette with deionized water.
2. Peel off the tape covering the slot on the back of the gel cassette.
3. In one fluid motion, pull the sample well former out of the cassette.

4. Use a 1-mL pipette to gently wash the cassette wells with 1X TBE electrophoresis buffer. Repeat twice, and then fill the sample wells with electrophoresis buffer.
5. Assemble the gel apparatus. Fill the upper buffer chamber with 200 mL of the 1X TBE electrophoresis buffer. Ensure that the upper buffer chamber is not leaking. If the level of the electrophoresis buffer drops, the apparatus will need to be reassembled. Fill the lower buffer chamber with approximately 200 mL of electrophoresis buffer through the gap between the gel tension wedge and the back of the lower buffer chamber; align the lid on the buffer core. The lid can only be firmly seated if the (−) electrode is aligned over the banana plug on the right.
6. After the PCR reaction has been assembled, give the tubes a quick centrifugation and then add 6 μL of 10X Bromophenol Blue/Xylene Cyanol loading dye to each sample and the control tube.
7. Load ~18 μL of the no-template master mix control into a well on the left side of the gel.
8. In the next well, load 10 μL of the 25-bp DNA ladder (20 ng/μL). Leave a gap of approximately 5 wells, and carefully load the sample into as many wells as required, keeping in mind that one well holds a maximum volume of 25 μL.
9. Electrophorese the 8% TBE gel at 200 V for 35 min. The xylene cyanol corresponds to the 200-bp DNA marker, and it will have migrated approximately 1/2 of the length of the gel.
10. Prepare 2 tubes of 0.5 and 2 mL for shearing the gel slices for each sample. Make a hole through the bottom of 0.5-mL tubes with an 18-gauge needle. Place each 0.5-mL tube into a 2-mL tube.
11. Prepare fresh SybrGreenI DNA stain; 10 μL stock in 100 mL of 1x TBE. Minimize exposure to light.
12. Stop the gel electrophoresis after 35 min and dismantle the PAGE apparatus.
13. Using a clean tray, stain the gel for 5 min.
14. Scan the gel and save the image.
15. Lay the gel down on the dark reader. Using a brand new razor blade, carefully excise the ~92-bp band (sometimes it can be a range from 90 to 95 bp).
16. Cut the gel slice approximately 1 cm × 0.5 cm to fit into the gel-shearing tubes. Transfer the gel slices into the 0.5-mL tubes.

17. Clean scanner surface with dH_2O and remove any pieces of gel material.
18. Clean scanner area: discard dye and rinse trays, and discard SybrGreen waste and sharps material appropriately.
19. With the lids tailing (left of tube position), centrifuge at 12,000 rpm at 4°C for 3 min. The gel slices should shear through the holes and collect into the bottom of the 2-mL tubes.
20. For each gel slice that was sheared into a 2-mL tube, add ~200 μL of elution buffer (5:1, LoTE:7.5 M ammonium acetate). Ensure that all sheared gel pieces are covered with elution buffer. Add more buffer if needed.
21. Mix well by vortexing. Pulse-centrifuge.
22. Elute gel slurries at 4°C overnight.

3.7. Library Preparation, Day 7

Purpose: To purify the amplified cDNA products (miRNA band) from the gel slurry and quantify the purified products by Agilent DNA1000 chip assay.

3.7.1. Precipitation and Purification of the PCR Products

1. Put on a clean pair of gloves and disposable lab coat.
2. Wipe down the assigned workstation, pipettors, and small equipment.
3. Lay down a new benchcoat.
4. Preheat a heat block to 65°C.
5. Prechill a microcentrifuge to 4°C.
6. Retrieve ice and all required reagents.
7. Thaw all reagents; vortex, and pulse-spin.
8. Retrieve the gel slurries from the previous day's PAGE gel from 4°C.
9. Vortex and pulse-spin.
10. Heat the gel slurries at 65°C for 10 min in the preheated heat block.
11. Vortex the tubes, pulse-centrifuge, and transfer the gel slurry from each tube onto the top of a Spin-X filter column.
12. Centrifuge the sample through the Spin-X column into the collection tube at 12,000 rpm for 4 min at 4°C.
13. Check each Spin-X Column and ensure that all the buffer has eluted through the filter. Recentrifuge the tubes if there is still liquid trapped in the gel material.
14. Transfer the eluate from one sample into a single sterile 1.5-mL tube, up to a maximum total volume of 400 μL. The total volume of eluate will depend on the amount of

excised gel and the volume of added elution buffer; adjust the amount of ethanol used to precipitate the PCR products accordingly, and add the reagents to the eluate as follows:

- 400 μL Eluate
- 40 μL Sodium acetate (3 M)
- 3 μL Mussel glycogen (20 mg/mL)
- Ethanol (100% EtOH)

15. Vortex and pulse-centrifuge. Chill the tubes at -20°C for a minimum of 10 min.
16. Centrifuge at 14,000 rpm at 4°C for 30 min.
17. Wash the pellet with 1 mL of 70% EtOH by adding the EtOH solution and inverting the tube. Centrifuge at 14,000 rpm at 4°C for 2 min. Discard the supernatant as previously.
18. Repeat the 70% EtOH wash.
19. Pulse-centrifuge the sample tube and carefully remove any residual ethanol by using a P200 pipette tip first to remove the majority of the supernatant, then finally using a P10 pipette tip to remove the last trace of solution. Mark the outside bottom of the tube to better locate the pellet when resuspending in buffer.
20. Allow the tube to air-dry for approximately 5–10 min at room temperature, until the white precipitate is no longer visible.
21. Resuspend each sample in a total volume of 12 μL Qiagen EB buffer.
22. If the purified PCR product was EtOH precipitated in more than one 1.5-mL tube, then pool the aliquots belonging to a single sample into one tube containing 12 μL. This is the library stock.
23. Perform the Agilent DNA 1000 assay according to the manufacturer's protocol, loading a 1 μL aliquot of the library stock onto the Agilent DNA 1000 chip.
24. Remove an aliquot of the library stock sufficient to make a final concentration of 15 nM dilution of the library for Illumina sequencing. Calculate the amount required for the dilution using the result for the molarity of the sample determined by the Agilent DNA 1000 assay. Dilute using Qiagen EB buffer supplemented with 0.1% Tween-20. In the case of a very dilute sample, no dilution of the stock may be necessary. Keep in mind that a minimum of 4 μL of 15 nM material is required for cluster generation.

3.8. Analysis of microRNA Sequence Data

Raw Illumina sequence data (seq.txt or qseq files) should first be reduced into a set of unique reads. If your sequencing experiment contained a pool of indexed libraries, the reads should first be separated by pool (based on barcode sequence). Unique sequence reads can then be aligned using Novoalign (www.novocraft.com), which can also trim the 3′ linker sequence from reads prior to alignment. Novoalign identifies all the equally highest scoring ungapped alignments in the genome allowing for multiple mismatches. For libraries that are processed in-house, all alignments are stored as 'bam' (binary sequence alignment/map) files and scripts for annotating the alignments and quantifying microRNAs can be found on our website (http://www.bcgsc.ca/platform/bioinfo/software). The number of reads representing each trimmed/mapped read (tag) can be converted into counts such that each aligned sequence is associated with a measure of abundance (the number of copies of that read in the flowcell). Individual tags can then be annotated based on their overlap with known miRNA genes, other ncRNA genes, or protein-coding genes. Tags overlapping known miRNAs (3) may not always comprise the exact sequence of the parent miRNA gene due to the presence of isomiRs resulting from untemplated 3′ additions or variability in cleavage by Drosha and Dicer (2). In any

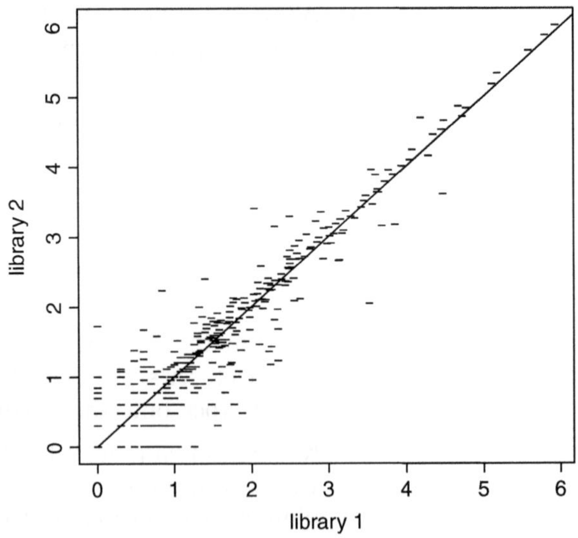

Fig. 14.2. Comparison of two microRNA biological replicate libraries for reproducibility. The reproducibility of any method of measuring microRNA expression is paramount for its utility in most applications. Shown are the microRNA expression values from two libraries prepared from the same RNA (from a human oligodendroglioma brain cancer). The scale is the base-ten logarithm of the expression values; hence, the top five miRNAs (all members of the let-7 family) produced between 150,000 and 1,000,000 individual reads. The Spearman correlation coefficient for these two libraries is 0.8935. The correlation for technical replicates (i.e., same library sequenced twice) is generally 0.99.

case, all unambiguously aligned tags for a miRNA can be considered for expression measurement whereas intronic and intergenic tags not corresponding to known miRNAs can be analyzed separately. A total summary expression value can be computed for each microRNA using the sum of all unambiguous (i.e., unique to a single miRNA) isomiR tags representing that microRNA. These individual expression values should be scaled (i.e., normalized) to compensate for variable sampling depth across sequencing runs. In a recent comparison of biological replicates, we have found that these expression values show very good reproducibility (**Fig. 14.2**). Some isomiRs result from enzymatic modification of the pre- or mature miRNA, resulting in nucleotides that do not match the corresponding position in the genome. These result in consistent mismatches visible in Novoalign alignments that can be readily retrieved from our database. Polymorphisms (and presumably mutations) would also produce similar features and hence these events can be identified routinely using this procedure. The tags aligning to introns or unannotated regions of the genome are candidates for novel microRNA genes. As previously described, the position and local genomic folding potential of these sites can reveal new conserved and nonconserved microRNA genes (2, 8).

4. Notes

1. This procedure was based heavily upon the Illumina Small RNA Sample Prep Protocol v 1.4B (www.illumina.com).
2. Nitrile or latex gloves should be worn at all times during these procedures.
3. When excising regions from a gel, it is acceptable (and is often the case) that there is no prominent band in the region of the gel specified.
4. Alternative 5′ adapters with indexing nucleotides (i.e., barcodes) may be used for preparing libraries that are to be sequenced as pools. Libraries constructed using individual indexed primers can be pooled in equal molarity for sequencing in a single lane.
5. The bromophenol blue dye will comigrate with the 20-nucleotide DNA marker.
6. Always use cold water circulation when electrophoresing gels.
7. When resuspending your pellet in EB buffer (**Section 3.7.1**), the volume may be decreased if the PCR product

yield is low as determined by the intensity of staining on the previous day's gel.

8. As the standard, Illumina miRNA library starts with gel excision of 14–30 nt RNA. Total RNA extracted using a column-based method may not be suitable; TRIZOL usually works well. RNA quality can be assessed using Agilent Bioanalyzer RNA Nano Chip to show acceptable RIN and Small RNA ChIP Assay to show small RNA content.
9. Dedicated work stations, solutions, and equipment free from nucleases are required, any contamination especially RNases may cause library failure.
10. A very practical exercise in the lab is that only a third of the ligation template is used in the initial PCR, to allow a repeat of the PCR or set up a different PCR thermal cycling protocol as opposed to repeating the entire procedure, especially if the starting material is limiting.

Acknowledgments

This work was funded, in part, by the BC Cancer Agency. The authors would like to thank all those (GSC and Illumina) who were involved in the microRNA library protocol development and Andy Chu for ongoing development of the data analysis pipeline.

References

1. O'Toole, A. S., Miller, S., Haines, N., Zink, M. C., and Serra, M. J. (2006) Comprehensive thermodynamic analysis of 3′ double-nucleotide overhangs neighboring Watson-Crick terminal base pairs. *Nucleic Acids Res* **34**, 3338–3344.
2. Morin, R. D., O'Connor, M. D., Griffith, M., Kuchenbauer, F., Delaney, A., Prabhu, A. L., Zhao, Y., McDonald, H., Zeng, T., Hirst, M., Eaves, C. J., and Marra, M. A. (2008) Application of massively parallel sequencing to microRNA profiling and discovery in human embryonic stem cells. *Genome Res* **18**, 610–621.
3. Griffiths-Jones, S., Saini, H. K., van Dongen, S., and Enright, A. J. (2008) miRBase: tools for microRNA genomics. *Nucleic Acids Res* **36**, D154–D158.
4. Kuchenbauer, F., Morin, R. D., Argiropoulos, B., Petriv, O. I., Griffith, M., Heuser, M., Yung, E., Piper, J., Delaney, A., Prabhu, A. L., Zhao, Y., McDonald, H., Zeng, T., Hirst, M., Hansen, C. L., Marra, M. A., and Humphries, R. K. (2008) In-depth characterization of the microRNA transcriptome in a leukemia progression model. *Genome Res* **18**, 1787–1797.
5. Bentley, D. R., Balasubramanian, S., Swerdlow, H. P., Smith, G. P., Milton, J., Brown, C. G., Hall, K. P., Evers, D. J., Barnes, C. L., Bignell, H. R., Boutell, J. M., Bryant, J., Carter, R. J., Keira Cheetham, R., Cox, A. J., Ellis, D. J., Flatbush, M. R., Gormley, N. A., Humphray, S. J., Irving, L. J., Karbelashvili, M. S., Kirk, S. M., Li, H., Liu, X., Maisinger, K. S., Murray, L. J., Obradovic, B., Ost, T., Parkinson, M. L., Pratt, M. R., Rasolonjatovo, I. M., Reed, M. T., Rigatti, R., Rodighiero, C., Ross, M. T., Sabot, A., Sankar, S. V., Scally, A., Schroth, G. P., Smith, M. E., Smith, V. P., Spiridou, A., Torrance, P. E., Tzonev, S. S., Vermaas, E. H.,

Walter, K., Wu, X., Zhang, L., Alam, M. D., Anastasi, C., Aniebo, I. C., Bailey, D. M., Bancarz, I. R., Banerjee, S., Barbour, S. G., Baybayan, P. A., Benoit, V. A., Benson, K. F., Bevis, C., Black, P. J., Boodhun, A., Brennan, J. S., Bridgham, J. A., Brown, R. C., Brown, A. A., Buermann, D. H., Bundu, A. A., Burrows, J. C., Carter, N. P., Castillo, N., Chiara, E. C. M., Chang, S., Neil Cooley, R., Crake, N. R., Dada, O. O., Diakoumakos, K. D., Dominguez-Fernandez, B., Earnshaw, D. J., Egbujor, U. C., Elmore, D. W., Etchin, S. S., Ewan, M. R., Fedurco, M., Fraser, L. J., Fuentes Fajardo, K. V., Scott Furey, W., George, D., Gietzen, K. J., Goddard, C. P., Golda, G. S., Granieri, P. A., Green, D. E., Gustafson, D. L., Hansen, N. F., Harnish, K., Haudenschild, C. D., Heyer, N. I., Hims, M. M., Ho, J. T., Horgan, A. M., Hoschler, K., Hurwitz, S., Ivanov, D. V., Johnson, M. Q., James, T., Huw Jones, T. A., Kang, G. D., Kerelska, T. H., Kersey, A. D., Khrebtukova, I., Kindwall, A. P., Kingsbury, Z., Kokko-Gonzales, P. I., Kumar, A., Laurent, M. A., Lawley, C. T., Lee, S. E., Lee, X., Liao, A. K., Loch, J. A., Lok, M., Luo, S., Mammen, R. M., Martin, J. W., McCauley, P. G., McNitt, P., Mehta, P., Moon, K. W., Mullens, J. W., Newington, T., Ning, Z., Ling Ng, B., Novo, S. M., O'Neill, M. J., Osborne, M. A., Osnowski, A., Ostadan, O., Paraschos, L. L., Pickering, L., Pike, A. C., Chris Pinkard, D., Pliskin, D. P., Podhasky, J., Quijano, V. J., Raczy, C., Rae, V. H., Rawlings, S. R., Chiva Rodriguez, A., Roe, P. M., Rogers, J., Rogert Bacigalupo, M. C., Romanov, N., Romieu, A., Roth, R. K., Rourke, N. J., Ruediger, S. T., Rusman, E., Sanches-Kuiper, R. M., Schenker, M. R., Seoane, J. M., Shaw, R. J., Shiver, M. K., Short, S. W., Sizto, N. L., Sluis, J. P., Smith, M. A., Ernest Sohna Sohna, J., Spence, E. J., Stevens, K., Sutton, N., Szajkowski, L., Tregidgo, C. L., Turcatti, G., Vandevondele, S., Verhovsky, Y., Virk, S. M., Wakelin, S., Walcott, G. C., Wang, J., Worsley, G. J., Yan, J., Yau, L., Zuerlein, M., Mullikin, J. C., Hurles, M. E., McCooke, N. J., West, J. S., Oaks, F. L., Lundberg, P. L., Klenerman, D., Durbin, R., and Smith, A. J. (2008) Accurate whole human genome sequencing using reversible terminator chemistry. *Nature* **456**, 53–59.

6. Reid, J. G., Nagaraja, A. K., Lynn, F. C., Drabek, R. B., Muzny, D. M., Shaw, C. A., Weiss, M. K., Naghavi, A. O., Khan, M., Zhu, H., Tennakoon, J., Gunaratne, G. H., Corry, D. B., Miller, J., McManus, M. T., German, M. S., Gibbs, R. A., Matzuk, M. M., and Gunaratne, P. H. (2008) Mouse let-7 miRNA populations exhibit RNA editing that is constrained in the 5′-seed/ cleavage/anchor regions and stabilize predicted mmu-let-7a:mRNA duplexes. *Genome Res* **18**, 1571–1581.
7. Li, H., and Durbin, R. (2000) Fast and accurate short read alignment with Burrows-Wheeler transform. *Bioinformatics* **25**(14), 1754–1760.
8. Friedlander, M. R., Chen, W., Adamidi, C., Maaskola, J., Einspanier, R., Knespel, S., and Rajewsky, N. (2008) Discovering microRNAs from deep sequencing data using miRDeep. *Nat Biotechnol* **26**, 407–415.

Chapter 15

Assessing In Vivo MicroRNA Function in the Germline Stem Cells of the *Drosophila* Ovary

Kin Chan and Hannele Ruohola-Baker

Abstract

A more complete understanding of the biology of adult stem cells could yield important insights toward devising effective cell-based regenerative therapies to treat disease. The germline stem cells (GSCs) in the fruit fly *Drosophila melanogaster* are an excellent in vivo model for the study of adult stem cell biology. There is increasing evidence from a growing field that microRNAs (miRNAs) play important roles in controlling many aspects of stem-cell biology. Using straightforward genetic manipulations combined with well-established cell biological analysis techniques, we and others have found that the miRNA pathway regulates the cell division rate of *Drosophila* GSCs as well as the maintenance of the GSCs in their niche. In this chapter, we offer a detailed, self-contained description of a general method to assess the in vivo functions of miRNAs in the GSCs of the *Drosophila* ovary.

Key words: microRNA, germline stem cell, GSC, *Drosophila*, dicer, adult stem cell, stem-cell division, stem-cell maintenance, stem-cell niche, confocal immunofluorescence.

1. Introduction

Adult stem cells are required to maintain many, if not all, tissues in metazoans. Such cells engage in asymmetric cell divisions, giving rise to another stem cell as well as a daughter cell that adopts a more differentiated fate (1). Thus, adult stem cells are self-renewing sources of differentiated cells and therefore, potential sources of cells for cell-based regenerative therapies to treat disease. However, significant challenges remain before such cells can be successfully used for regenerative medicine. Among these challenges is the need to devise the means to control the

B. Zhang, E.J. Stellwag (eds.), *RNAi and MicroRNA-Mediated Gene Regulation in Stem Cells*, Methods in Molecular Biology 650, DOI 10.1007/978-1-60761-769-3_15, © Springer Science+Business Media, LLC 2010

microenvironment into which to transplant the stem cells, as the microenvironment has important effects on the fates of these cells (*see* (2) for a recent review). A comprehensive, broad-based understanding of adult stem-cell biology in an in vivo context is likely to yield insights to aid in the design of cell-based therapies. In addition, the study of adult stem-cell biology might elucidate aspects of cancer stem-cell biology, with possible implications for the treatment of various cancers (3).

Control of adult stem-cell division and maintenance are both inextricably connected to the function of these cells. The proper regulation of adult stem-cell division is crucial, in order to avoid either excessive or insufficient cell division rates (which could lead to tumorigenesis or tissue atrophy, respectively). Similarly, mechanisms for regulating stem-cell maintenance must exist in order to ensure that tissues retain sufficient regenerative capacity throughout the lifetime of an animal. The *Drosophila melanogaster* germline stem cell (GSC) system is an excellent in vivo model for the investigation of these two related processes (4). The GSCs reside in the anterior part of each ovariole, which is referred to as the germarium. The GSCs are closely apposed to the cap cells, which are key to defining the stem-cell niche, although there is evidence to suggest that escort stem cells and escort cells probably also play roles in niche function (5). Consistent with the self-renewing cell-division paradigm of adult stem cells, GSCs divide asymmetrically to yield another GSC and a cystoblast. The resulting GSC is retained in the niche while the cystoblast undergoes further differentiation (6) (*see* **Fig. 15.1**).

Using a combination of straightforward genetic manipulations and cell biological analysis, we and other labs have determined that the microRNA (miRNA) pathway regulates both GSC division and maintenance, in a cell-autonomous manner (7–13). Thus far, only a handful of microRNAs have been shown to regulate GSC biology in vivo. For example, the *bantam* miRNA is required for GSC maintenance (11, 14, 15). Similarly, loss of *miR-278* results in a slower GSC division rate while loss of *mir-7* results in abnormally high expression of Cyclin E in GSCs, also suggestive of abnormal cell-cycle progression (13). *miR-278* and *miR-7* bind the 3′ untranslated region (UTR) of the *dap* (*dacapo*) mRNA (13). *dap* encodes the sole cyclin-dependent kinase inhibitor that regulates the G_1/S transition in *Drosophila* (16, 17). Moreover, ablation of the miRNA pathway in GSCs by loss of Dicer-1 (Dcr-1) function causes a delay in the G_1/S transition (8). Consistent with this, *dcr-1*^{Q1147X} GSCs exhibit derepression of reporter constructs bearing either 630 or 866 bp of the *dap* 3′ UTR (13). Taken together, these findings suggest that multiple miRNAs (including *miR278* and *miR-7*) regulate cell divisions in GSCs by controlling the expression of *dap*.

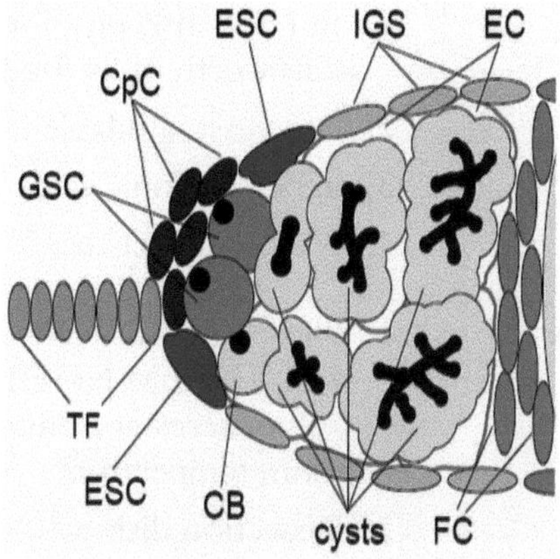

Fig. 15.1. Schematic diagram of a germarium. Germline stem cells (GSCs) are located near the anterior tip, in close apposition to the cap cells (CpCs), which define the stem-cell niche. An organelle called the spectrosome is oriented toward the cap cells. Spectrosomes and their descendant fusomes (*shown in black*) can be visualized by staining with an antibody against adducin. Asymmetric division by a GSC gives rise to another GSC and a cystoblast (CB). The cystoblast continues to divide and differentiate, successively resulting in 2-, 4-, 8-, and 16-cell cysts. The other cell types shown are terminal filament (TF), inner germarial sheath (IGS), escort stem cells (ESC), escort cells (EC), and follicle cells (FC).

Many further investigations will be required to elucidate the in vivo functions of other miRNAs in GSC biology. Such work is likely to yield useful insights into the biology of adult stem cells, in general. The remainder of this chapter describes a broadly applicable experimental procedure to carry out such investigations. Germline clones are generated using the yeast-derived heat shock-inducible flipase/flipase recombination target (hsFLP/FRT) system (18). Clonal induction at different life stages can have very different effects on, and yield more nuanced insight into, GSC biology (*see* **Section 3**). GSCs with the desired genotypes are stained using antibodies against markers of interest. The germline tissue then is examined using confocal immunofluorescence to quantify GSC division and maintenance rates. We also describe a variation of this procedure to detect expression of miRNAs using sensor constructs.

2. Materials

2.1. For Clonal Induction

1. Fly food: Standard Bloomington *Drosophila* medium consisting of 39 L water, 675 g dry yeast, 390 g soy flour, 2.85 kg yellow cornmeal, 225 g agar, 3 L light corn syrup,

and 188 mL propionic acid. These quantities yield >4000 vials worth of fly food.

2. Dry yeast, available from grocery store. Store at 4°C.
3. 0.5% Propionic acid. Store at room temperature.
4. Water bath set to 37°C.

2.2. For Dissection of Ovaries

1. Phosphate-buffered saline (PBS) 10X: 1.37 M NaCl, 27 mM KCl, 100 mM Na_2HPO_4, 18 mM KH_2PO_4, adjusted to pH 7.4 as necessary. Autoclave or sterile filter before storage at room temperature.
2. Dissection dish.
3. Two pairs of forceps (e.g., model 17-305 from Miltex, York, PA).
4. Transfer pipets (e.g., BD Falcon 357524).

2.3. For Antibody Staining

1. 5% Paraformaldehyde in 1X PBS: Make fresh by mixing the following according to these proportions: 3.125 mL of 16% paraformaldehyde stock with 5.875 mL of ddH_2O and 1 mL of 10X PBS.
2. PBT: 0.2% Triton X-100 in 1X PBS. Store at room temperature.
3. PBTB: 0.2% bovine serum albumin, 5% normal goat serum, balance PBT. Store at 4°C.
4. Mounting medium: Mix 800 mL of 80% glycerol, 3% *n*-propyl gallate with 200 mL of Prolong Gold reagent (Invitrogen, Carlsbad, CA). Store at 4°C.
5. 100X DAPI (4′,6-diamidino-2-phenylindole): 0.1 mg/mL in ddH_2O, store at 4°C.
6. Anti-adducin antibody 1B1 (mouse IgG1) from Developmental Studies Hybridoma Bank.
7. Anti-lamin C antibody LC28.26 (mouse IgG1) from Developmental Studies Hybridoma Bank.
8. Other primary antibody of interest to the investigator, raised in a nonmouse host species.
9. Antimouse secondary antibody coupled to a red fluor (e.g., Alexa Fluor 568).
10. An appropriate secondary antibody against the nonmouse-derived primary antibody (Item 8 in this list), coupled to a far red fluor (e.g., Alexa Fluor 647).
11. Precleaned glass slides.
12. 50 × 24 mm cover slips.
13. Two 22-gauge syringe needles or similar.

2.4. For Sensor Construct Analysis

1. 5% Paraformaldehyde in 1X PBS: Make fresh by mixing the following according to these proportions: 3.125 mL of 16% paraformaldehyde stock with 5.875 mL ddH_2O and 1 mL 10X PBS.
2. PBT: 0.2% Triton X-100 in 1X PBS. Store at room temperature.
3. PBTB: 0.2% bovine serum albumin, 5% normal goat serum, balance PBT. Store at 4°C.
4. Mounting medium: Mix 800 mL of 80% glycerol, 3% *n*-propyl gallate with 200 mL of Prolong Gold reagent (Invitrogen, Carlsbad, CA). Store at 4°C.
5. Antiadducin antibody 1B1 (mouse IgG1) from Developmental Studies Hybridoma Bank.
6. Antilamin C antibody LC28.26 (mouse IgG1) from Developmental Studies Hybridoma Bank.
7. Antimouse secondary antibody coupled to a red fluor (e.g., Alexa Fluor 568).
8. Precleaned glass slides.
9. 50 × 24-mm cover slips.
10. Two 22-gauge syringe needles or similar.

3. Methods

To generate mosaic germline tissue, one must first obtain larvae of a genotype that is susceptible to hsFLP/FRT-mediated mitotic recombination. For example, the genotype for generating clones lacking the miRNA *bantam* would be: *hsFLP;;ban FRT80B/Ubi-GFP FRT80B*. The chromosome bearing a wild-type copy of *bantam* is marked with GFP. Thus, any clonal GSC (and its descendant cells) would have the genotype *hsFLP;;ban FRT80B* and be distinguishable by a *lack* of GFP. Flies with a wild-type copy of the gene of interest on an unmarked FRT-bearing chromosome would serve as an appropriate control. In this example, such control flies would have the genotype: *hsFLP;;FRT80B/Ubi-GFP FRT80B*. To maximize consistency, controls should be treated and processed in parallel with mutant samples. The theory and means of obtaining flies of a genotype of interest are beyond the scope of this chapter.

To detect the expression of miRNAs in situ, we use flies that bear a sensor construct transgene of interest. Such constructs can be classified into two categories: miRNA sensors and 3′ UTR sensors (*see* **Fig. 15.2**). A typical miRNA sensor consists of two target

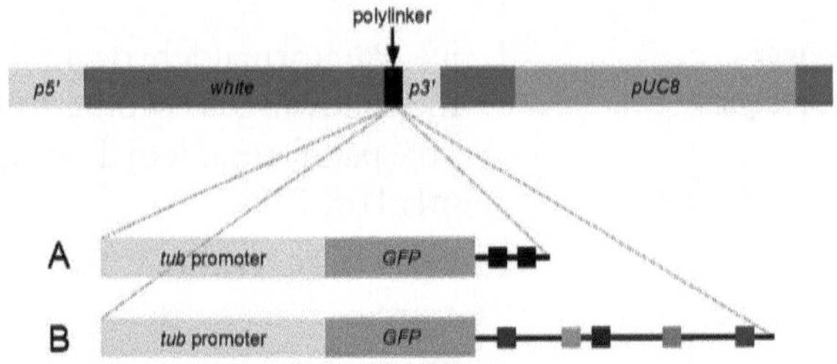

Fig. 15.2. Structures of miRNA and 3′ UTR sensor constructs. (**a**) The structure of an miRNA sensor construct. Two copies of an miRNA target site of interest (shown as *filled squares*) are inserted downstream of GFP under the control of the tubulin promoter. Each target site is a perfect complement to its cognate miRNA. (**b**) The structure of a 3′UTR sensor. The 3′ UTR of a gene of interest is ligated downstream of tub-GFP. The various miRNA target sites in the 3′ UTR are represented as squares with different shading. Also, since the target sites in a 3′ UTR sensor are naturally occurring, each target site will not be perfectly complementary to its cognate miRNA. In either case, the construct is ligated into the polylinker of a vector such as the *P* element of CaSpeR4 (depicted at the *top* of figure), which is then injected into flies to produce transgenic lines.

sequences that are each a perfect complement for an miRNA of interest. These target sequences are ligated downstream of a reporter such as *tub-GFP*. This construct is then ligated into the polylinker of a vector such as the *P* element in CaSpeR4, which is then injected into flies to generate transgenic lines. On the other hand, a 3′ UTR sensor consists of a portion of the 3′ UTR of a gene of interest that is fused downstream of a *tub-GFP* reporter. In contrast to the target sites used in an miRNA sensor, the naturally occurring target sites used in a 3′ UTR sensor will not be perfectly complementary to their respective cognate miRNAs. A microRNA sensor is used to detect the expression of an miRNA of interest in GSCs, while a 3′ UTR sensor is used to determine whether the predicted miRNA sites in the 3′ UTR in question are functional in GSCs.

In either case, the function of cognate miRNA(s) in GSCs would result in silencing (or reduction) of the reporter while lack of expression would result in expression of the reporter. Clonal GSCs of interest can be generated in a sensor-bearing genetic background. This can be an especially useful technique, as miRNA expression can be compared side-by-side between mutant and nonmutant GSCs within the same germarium. For example, miRNA sensors should be more abundant in *dcr-1* mutant GSCs compared to wild-type GSCs if the miRNAs in question are truly expressed in GSCs. Since GFP is used to monitor the reporter's expression, the appropriate FRT-bearing chromosomes should be marked by a different reporter, such as *armadillo-lacZ*. Again, the details of the theory and means of constructing these sensors and of obtaining appropriately integrated fly lines are beyond the scope of this chapter.

Care should be taken to ensure that enough flies of interest are generated to accommodate damaged tissue and variability in generating and analyzing clonal germaria. It is wise to allot a minimum of seven ovaries for each antibody-staining preparation, as tissue can be lost or damaged during the multiple pipeting steps. Starting with too few ovaries can impact the number of informative germaria from any preparation, a potentially important issue since the yield of clonal germaria using the hsFLP/FRT system can be highly variable depending on the genotype being generated and the FRT being used. Also, note that ovary dissections on more than 1 day will be necessary to assess GSC maintenance in the niche.

Proper handling of the flies is crucial to the success of this procedure. After eclosion, females of the correct genotype are maintained on a rich diet that includes wet yeast. An equal number of males are included in each vial. The flies should be transferred to fresh vials with wet yeast every 2–3 days after eclosion. These steps are necessary to encourage expansion of the ovarian tissue, which significantly facilitates subsequent steps. Flies should be kept in a 25°C incubator when not being handled.

Finally, due consideration should be given to the timing of clonal induction. For example, induction of *dcr-1*^{Q1147X} clones during late larval/early pupal development causes a slower rate of GSC division, while not adversely affecting GSC maintenance (8, 11). In contrast, induction of such clones in adulthood results in both GSC division and maintenance defects (9, 11). Similarly, induction of *Mad*^{12} clones in the preadult stages does not result in discernible GSC division or maintenance phenotypes while adult induction causes defects in both processes (11). These observations show that clonal induction in adults can result in different phenotypes compared to induction in the preadult stages. In the case of the *dcr-1*^{Q1147X}, control over the processes of GSC division and maintenance became genetically separable, but only if clones were induced at a preadult stage.

3.1. Clonal Induction

3.1.1. Clonal Induction in Late Larvae/Early Pupae

1. Set up crosses using five virgins and five males (of the appropriate genotypes) per vial. Assuming the fly strains are healthy, a minimum of three vials per cross is recommended. This number should be adjusted upwards as necessary for sickly strains (*see* **Note 1**).
2. After 3 days, transfer the adults to fresh vials (*see* **Note 2**).
3. At 6–7 days after the initial setup of the cross, heat shock the third instar larvae/early pupae for 60 min in the 37°C water bath (*see* **Note 3**).

4. On the following day, heat shock the larvae/pupae for 60 min at 37°C.
5. At 2 days after flies start to eclose, collect all (adult) females of the desired genotype into a common pool. Redistribute ~10 females into each vial with an equal number of males (of any genotype). Each vial should include a dollop of wet yeast (*see* **Notes 4** and **5**).
6. Transfer the flies to fresh vials with wet yeast every 2–3 days until the completion of the experiment. Keep flies at 25°C (*see* **Note 6**).

3.1.2. Clonal Induction in Adults

1. Set up crosses using five virgins and five males (of the appropriate genotypes) per vial (*see* **Note 1**).
2. After 3 days, transfer the adults to fresh vials (*see* **Note 2**).
3. At 2–4 days after flies start to eclose, collect all (adult) females of the desired genotype into an empty vial. Heat shock at 37°C for 50 min (*see* **Notes 3** and **5**).
4. Following heat shock, distribute the females into vials (with food) supplemented with wet yeast. Place ~10 females into each vial with the same number of males (*see* **Note 4**).
5. On the day following the first heat-shock treatment, collect the females into an empty vial and heat shock again at 37°C for 50 min.
6. Following the second heat-shock treatment, redistribute the females into fresh vials with wet yeast, as in step 4.
7. Transfer the flies to fresh vials with wet yeast every 2–3 days until the completion of the experiment. Keep flies at 25°C (*see* **Note 6**).

3.2. Dissection of Ovaries

1. Anesthetize the required number of females using CO_2.
2. Fill the dissecting dish with 1X PBS.
3. Using one of the forceps, grasp a fly by the thorax and immerse into the dish.
4. Using the other forceps, grasp the posterior tip of the abdomen and rip the posterior tip away (*see* **Notes 7** and **8**).
5. Using the arms of the forceps used in step 4, gently squeeze the abdomen to push the ovaries into the media through the opening created in step 4 (*see* **Note 9**).
6. Repeat steps 3–5 until all ovaries have been collected (*see* **Note 10**).
7. Using a transfer pipet, distribute the ovaries into microfuge tubes so that there are at least 7 per tube for each antibody staining preparation (*see* **Note 11**).

3.3. Antibody Staining

1. On completion of dissections, remove excess PBS and fix the ovaries in the 5% paraformaldehyde solution for 10 min, with nutating at room temperature (*see* **Notes 12** and **13**).
2. After fixing, wash the ovaries four times, using 1 mL of PBT and nutating for 10–15 min each time at room temperature.
3. Block the ovaries with 0.5 mL of PBTB for 1 h at room temperature (*see* **Note 14**).
4. Remove the PBTB used for blocking.
5. Add primary antibodies, diluted appropriately in PBTB. We recommend using the 1B1 and LC28.26 antibodies (against adducin and lamin C, respectively) at 1:20. Incubate overnight while nutating at 4°C (*see* **Notes 15** and **16**).
6. Wash ovaries as in step 2.
7. Add secondary antibody or antibodies, diluted appropriately in PBTB. Incubate while nutating at 4°C overnight or at room temperature for 2–3 h.
8. Carry out washes as described in step 2, but add 10 mL of 100X DAPI to the second wash.
9. After removing excess liquid, add at least 50 mL of mounting medium (*see* **Note 17**).
10. Pipet the ovaries onto a clean glass slide. Using the 22-gauge needles, tease the ovarioles apart from one another (*see* **Note 18**).
11. Apply a cover slip over the sample to achieve even spreading of the mounting medium (*see* **Note 19**).

3.4. Sensor Construct Analysis

To prepare samples for sensor construct analysis, procedure 3.3 can be readily applied. If one is able to identify the GSCs solely by morphology and if one is only interested in observing sensor expression as a function of GSC genotype, staining with the 1B1 and LC28.26 antibodies can be omitted. However, staining with these antibodies is recommended to ensure unambiguous identification of the GSCs.

4. Notes

1. Our lab typically supplements vials used for standard fly husbandry with a thin layer of dry yeast.

2. After 3 days, there are usually enough eggs laid within each vial that a productive experiment can be carried out. However, to maximize the payoff from the fly husbandry work required to set up an experiment, we transfer these P_0 flies to fresh vials on the third day and again on the sixth day to obtain two more rounds of egglaying.
3. We have found that it is optimal to carry out the first heat shock on the F_1's derived from the first 3-day round of egglaying at day 7 after cross setup. For F_1's derived from subsequent rounds of egglaying, the first heat shock is carried out on day 6 after transferring the P_0's to fresh vials. This is because there is a relative delay in efficient progeny production upon the initial setup of the cross.
4. Wet yeast is prepared by mixing dry yeast with 0.5% propionic acid until a paste that is the consistency of peanut butter is formed. We apply a smear of wet yeast to the inner wall of each vial using a spatula.
5. One can also be more fastidious about collecting flies at this stage by sorting for females of the desired genotype on a daily basis as soon as flies eclose. If the P_0's are particularly prolific, it would be necessary to sort daily in order to avoid excessive crowding of the F_1's.
6. If assessing GSC maintenance, it becomes necessary to sample from multiple time points; collecting ovaries at days 7, 14, and 21 after the second heat shock treatment is a reasonable start. Empirical observation should guide the choice of a more optimal series of time points.
7. Debris from dissecting should be wiped off onto a wet paper towel, as debris in the dissecting dish can obscure vision.
8. Sometimes, the abdomen will become separated from the thorax. Some care in positioning the grasping forceps to grip the anterior-most part of the abdomen can help to prevent this. In case of abdomen separation, hold on to the posterior end, re-establish a grasp of the anterior-most portion of the abdomen, and proceed as normal. Alternatively, hold onto the posterior end and use the other forceps to squeeze the contents of the abdomen out through the opening that had been attached to the thorax. Pins can also be used in the dissection process.
9. Use the arms of the forceps to squeeze the abdominal contents out. Do not use the tips, as these will easily rip the tissue.

10. Work expeditiously. As a general rule, avoid having any dissected ovaries sit for over 30 min before the beginning of fixation by paraformaldehyde.
11. Avoid drawing ovaries into the flared section of the transfer pipet, as those will typically remain permanently stuck there. Alternatively, ovaries can be pipetted using a P-1000 micropipetter.
12. A minimum of 250 μL is necessary to achieve mixing within each microcentrifuge tube while rocking on a nutator. If necessary, manually roll the tubes to establish adequate flowing of the fixing solution before putting on nutator.
13. To maximize consistency, use aliquots of the same batches of solutions for both experimental and control samples at each step. Also, process experimental and control samples in parallel as much as possible.
14. Ovaries can also be blocked overnight at 4°C.
15. The LC28.26 antibody stains the cytoskeleton on the cap cells, which form the stem-cell niche. The 1B1 antibody stains the spectrosome, an organelle that is specific to the GSC and is oriented toward the cap cells. As the combination of these two markers should enable even the novice to unambiguously identify the GSCs, staining using these antibodies is recommended.
16. Incubation in primary and secondary antibodies can be done using as little as 200 mL of solution per microcentrifuge tube.
17. After completion of staining, ovaries should be stored in mounting medium at least overnight before slide preparation. Immediate slide preparation (before adequate equilibration with the mounting medium) can result in distortion and damage to the tissue.
18. Use the needles to separate the late-stage eggs from one another. This should result in clean separation of the attached ovarioles from one another as well. Avoid touching the germaria themselves. With some practice, it is possible to arrange all of the ovarioles from an ovary in a neat row or column. Such an arrangement can greatly facilitate cell biological analysis.
19. Some of the mounting medium should be applied onto the cover slip before placement over the glass slide. If necessary, additional mounting medium can be applied to the edges of the cover slip to enable even spreading of the medium under the entirety of the cover slip.

Acknowledgment

The authors thank Karin Fischer for her critical reading of the manuscript. This work was supported by the grants from NIH for HRB.

References

1. Fuchs, E., Tumbar, T., and Guasch, G. (2004) Socializing with the neighbors: stem cells and their niche. *Cell* **116**, 769–778.
2. Ting, A. E., Mays, R. W., Frey, M. R., Hof, W. V., Medicetty, S., and Deans, R. (2008) Therapeutic pathways of adult stem cell repair. *Crit Rev Oncol Hematol* **65**, 81–93.
3. Yang, Y. M., and Chang, J. W. (2008) Current status and issues in cancer stem cell study. *Cancer Invest* **26**, 741–755.
4. Fuller, M. T., and Spradling, A. C. (2007) Male and female *Drosophila* germline stem cells: two versions of immortality. *Science.* **316**, 402–404.
5. Decotto, E., and Spradling, A. C. (2005) The *Drosophila* ovarian and testis stem cell niches: Similar somatic stem cells and signals. *Dev Cell* **9**, 501–510.
6. Dansereau, D. A., and Lasko, P. (2008) The development of germline stem cells in *Drosophila*. *Methods Mol Biol* **450**, 4503–4526.
7. Förstemann, K., Tomari, Y., Du, T., Vagin, V. V., Denli, A. M., Bratu, D. P., Klattenhoff, C., Theurkauf, W. E., and Zamore, P. D. (2005) Normal microRNA maturation and germ-line stem cell maintenance requires Loquacious, a double-stranded RNA-binding domain protein. *PLoS Biol* **3**, e236.
8. Hatfield, S. D., Shcherbata, H. R., Fischer, K. A., Nakahara, K., Carthew, R. W., and Ruohola-Baker, H. (2005) Stem cell division is regulated by the microRNA pathway. *Nature* **435,** 974–978.
9. Jin, Z., and Xie, T. (2007) Dcr-1 maintains *Drosophila* ovarian stem cells. *Curr Biol* **17**, 539–544.
10. Park, J. K., Liu, X., Strauss, T. J., McKearin, D. M., and Liu, Q. (2007) The miRNA pathway intrinsically controls self-renewal of *Drosophila* germline stem cells. *Curr Biol* **17**, 533–538.
11. Shcherbata, H. R., Ward, E. J., Fischer, K. A., Yu, J., Reynolds, S. H., Chen, C., Xu, P., Hay, B. A., and Ruohola-Baker, H. (2007) Stage-specific differences in the requirements for germline stem cell maintenance in the *Drosophila* ovary. *Cell Stem Cell* **1**, 698–709.
12. Yang, L., Chen, D., Duan, R., Xia, L., Wang, J., Qurashi, A., Jin, P., and Chen, D. (2007) Argonaute 1 regulates the fate of germline stem cells in *Drosophila*. *Development* **134**, 4265–4272.
13. Yu, J., Reynolds, S. H., Hatfield, S. D., Shcherbata, H. R., Fischer, K. A., Ward, E. J., Long, D., Ding, Y., and Ruohola-Baker, H. (2009) Dicer-1-dependent Dacapo suppression acts downstream of Insulin receptor in regulating cell division of *Drosophila* germline stem cells. *Development* **136**, 1497–1507.
14. Neumüller, R. A., Betschinger, J., Fischer, A., Bushati, N., Poernbacher, I., Mechtler, K., Cohen, S. M., and Knoblich, J. A. (2008) Mei-P26 regulates microRNAs and cell growth in the Drosophila ovarian stem cell lineage. *Nature* **454**, 241–245.
15. Yang, Y., Xu, S., Xia, L., Wang, J., Wen, S., Jin, P., and Chen, D. (2009) The bantam microRNA is associated with drosophila fragile X mental retardation protein and regulates the fate of germline stem cells. *PLoS Genet* **5**, e1000444.
16. de Nooij, J. C., Letendre, M. A., and Hariharan, I. K. (1996) A cyclin-dependent kinase inhibitor, Dacapo, is necessary for timely exit from the cell cycle during Drosophila embryogenesis. *Cell* **87**, 1237–1247.
17. Lane, M. E., Sauer, K., Wallace, K., Jan, Y. N., Lehner, C. F., and Vaessin, H. (1996) Dacapo, a cyclin-dependent kinase inhibitor, stops cell proliferation during Drosophila development. *Cell* **87**, 1225–1235.
18. Golic, K. G. and Lindquist, S. (1989) The FLP recombinase of yeast catalyzes site-specific recombination in the *Drosophila* genome. *Cell* **59**, 499–509.

Chapter 16

Monitoring MicroRNA Expression During Embryonic Stem-Cell Differentiation Using Quantitative Real-Time PCR (qRT-PCR)

Xiaoping Pan, Alexander K. Murashov, Edmund J. Stellwag, and Baohong Zhang

Abstract

Quantitative real-time PCR (qRT-PCR) is a reliable method to determine and monitor microRNA (miRNA) expression profiles in different cells, tissues, and organisms. Although there are several different strategies for performing qRT-PCR to determine miRNA expression, all of them have two steps in common: reverse transcription for obtaining cDNA from mature miRNA sequence and standard real-time PCR for amplification of cDNA. This chapter demonstrates the application of TaqMan-based real-time PCR for determining miRNA expression profiles during mouse embryonic stem-cell differentiation. In this method, a mature miRNA sequence is first reverse transcribed into a long cDNA with a 40- to 50-nt miRNA-specific stem-loop primer; then, a standard real-time PCR reaction is performed for determining miRNA expression using a forward miRNA-specific primer, a universal reverse primer, and FAM dye-labeled TaqMan probes.

Key words: Embryonic stem cell, microRNA, quantitative real-time PCR, qRT-PCR, differentiation, gene expression.

1. Introduction

MicroRNAs (miRNAs) are a newly discovered class of endogenous small RNAs that inhibit gene expression by binding to their mRNA targets through complementary base pairing and activation of an RNA degrading system or blockage of the protein translation machinery. The most common target sequences for miRNAs are located in the 3′ untranslated regions (UTR) of the target mRNAs but target sequences have also been identified

B. Zhang, E.J. Stellwag (eds.), *RNAi and MicroRNA-Mediated Gene Regulation in Stem Cells*, Methods in Molecular Biology 650, DOI 10.1007/978-1-60761-769-3_16,

in the 5′ UTR and in the protein-coding regions (1–3). Perfect complementary base-pair annealing between an miRNA and its mRNA target usually leads to the cleavage of the mRNA whereas imperfect base pairing results in inhibition of protein translation.

Currently, a total of 10,833 miRNAs have been identified and deposited in the miRNA database, miRBase (Release 14 at September 2009) (4); these miRNAs were obtained from 115 animal, plant, or virus species. Of them, 721, 579, and 325 miRNAs were obtained from human, mouse, and rat, respectively. Computational studies have shown that miRNAs potentially regulate more than 30% of the protein-coding genes in humans (5, 6). However, these estimates are based on limited data and as more miRNA are identified, this number will likely increase significantly. An increasing body of evidence suggests that miRNAs regulate gene expression as well as play important functions in almost all biological and metabolic processes, including signal transduction, development, disease, and response to environmental biotic and abiotic stresses (7–9). More pertinent to the theme of this book, recent studies have shown that miRNAs control stem-cell maintenance and differentiation (10–13). Several investigations have shown that a set of miRNAs are differentially expressed in stem cells compared to other tissues and that these miRNAs are also differentially expressed during stem-cell differentiation (10–30). Thus, investigating the expression profiles of miRNAs during stem-cell maintenance and differentiation are critical for understanding miRNA function in stem cells.

There are several existing approaches for determining miRNA expression from different tissues and species, including Northern blotting, microarray, quantitative real-time PCR (qRT-PCR), and most recently next-generation high-throughput sequencing. Each method has both advantages and disadvantages for determining the expression level of miRNAs. Currently, qRT-PCR has become one of the most powerful methods for determining miRNA expression in a variety of biological and metabolic processes. In this chapter, we will focus on a qRT-PCR protocol for monitoring miRNA expression during mouse embryonic stem-cell differentiation.

There are slight differences among different qRT-PCR protocols, with special considerations for those used to quantify miRNAs. Experimental analysis demonstrates that TaqMan-based real-time PCR quantitation of miRNAs is a reliable method for determining miRNA expression. In particular, the TaqMan-based system can be used to distinguish mature miRNAs from their precursor sequences as well as distinguish miRNAs in which there are only a single nucleotide differences between the miRNAs being compared (31). TaqMan-based qRT-PCR quantification of miRNAs includes two important steps: a reverse transcription reaction initiated by a miRNA-specific stem-loop structure, reverse transcription primer, followed by real-time quantitative PCR reaction (31). Because mature miRNA sequences are very short, predominately only about 20–24 nt in length, it is very critical to achieve complete and accurate amplification of the miRNA sequence, which is ensured by the miRNA-specific stem-loop reverse transcription primer. Then, the regular standard real-time PCR will be employed to amplify the cDNA and monitor the expression level of a selected miRNA. This method has been widely used by different laboratories, including ours for determining and monitoring miRNA expression in a variety of animal and plant species.

2. Materials

2.1. Cell Culture and Differentiation

1. Undifferentiated, pluripotent mouse embryonic stem (ES) D3 cell line was purchased from the American Type Culture Collection (Manassas, VA). This cell line can be cultured in nutrient medium or stored in liquid nitrogen.
2. Knock-Out Dulbecco's Modified Eagle's Medium (DMEM) (Invitrogen, Carlsbad, CA, Cat. No. 10829-018), for culturing of undifferentiated embryonic stem cells, supplemented with 15% fetal bovine serum (ES grade, Invitrogen, Carlsbad, CA, Cat. No. 16141-061).
3. Iscove's Modified Dulbecco's Medium (Invitrogen, Carlsbad, CA, Cat. No. 21056-023) supplemented with 15% fetal bovine serum (Invitrogen, Carlsbad, CA, Cat. No. 16000-036), used for differentiation of ES cells.
4. Neurobasal medium (Invitrogen, Carlsbad, CA, Cat. No. 21103-049) for maintenance of neuronal culture, supplemented with B27 (Invitrogen, Carlsbad, CA, Cat. No. 17504-044).
5. Penicillin–Streptomycin, liquid (Invitrogen, Carlsbad, CA, Cat. No. 15140-122). Use at a concentration of 100 U/mL of penicillin and 100 μg/mL of streptomycin. Store aliquoted at –20°C for indicated period of time.
6. L-glutamine, 200 mM (100X) (Invitrogen, Carlsbad, CA, Cat. No. 25030-081). Dilute 1:100 to a final concentration of 2 mM. Store aliquoted at –20°C for indicated period of time.
7. MEM nonessential amino acids, 100X solution (Invitrogen, Carlsbad, CA, Cat. No. 11140-050). Dilute 1:100 to a final concentration of 0.1 mM. Store at 4°C for indicated period of time.

8. 1400 U/mL murine leukemia inhibitory factor (Millipore, Billerica, MA, Cat. No. LIF1010). Dilute to a final concentration 1400 U/mL. Store at 4°C up to 12 months.
9. 2-Mercaptoethanol, 1000X, liquid (Invitrogen, Carlsbad, CA, Cat. No. 21985-023). Dilute 1:1000 to final concentration 55 μM.
10. Trans retinoic acid (Sigma, St. Lois, MO, Cat. No. R2625). Prepare concentrated stock solution as follows: dissolve 50 mg of retinoic acid in 78 mL of 100% ethanol and 4 mL distilled sterile water, 2×10^{-3} M. Then dilute 10 μL of that stock with 190 μL of 50% ethanol/water to 1×10^{-4} M. Dilute to final concentration of 1×10^{-6} M. Store stock solutions in the dark at 4°C.
11. Phosphate-buffered saline (PBS) (Invitrogen, Carlsbad, CA, Cat. No. 10010-023).
12. 0.05% Trypsin/EDTA, liquid (Invitrogen, Carlsbad, CA, Cat. No. 25300-054). Ready to use. Store aliquoted at −20°C for indicated period of time.
13. RNAlater (Ambion, Austin, TX). Store at room temperature.

2.2. MiRNA Isolation

1. *mir*Vana™ miRNA Isolation kit (Ambion, Austin, TX).
   a. MiRNA Wash solution 1. Before usage, add 21 mL of 100% ethanol. MiRNA Wash Solution 1 contains guanidinium thiocyanate, which is a potential biohazard and should be handled with caution.
   b. Wash solutions 2 and 3. Before the first usage, add 40 mL of 100% ethanol. This solution can be left at room temperature for up to 1 month. For longer storage periods, store at 4°C; but warm chilled solution up to room temperature before use.
   c. Collection tubes. Store at room temperature.
   d. Filter cartridges. Store at room temperature.
   e. Lysis/binding buffer. Store at 4°C.
   f. MiRNA homogenate additive. Store at 4°C.
   g. Acid-Phenol: Chloroform. Store at 4°C. Phenol is a poison and an irritant and therefore gloves or other protection should be worn when handling this reagent. Dispose phenol waste appropriately.
   h. Elution solution or nuclease-free water. Preheated to 95°C when used and stored at 4°C or room temperature.
2. 100% RNase-free ethanol stored at room temperature. Ethanol is flammable, so handle and dispose it accordingly.
3. RNase-free water.

2.3. Quantitative Real-Time PCR (qRT-PCR) Analysis

1. TaqMan® microRNA Reverse Transcription Kit (Applied Biosystems, Foster City, CA). All components should be stored at −20°C. All contents should be thawed on ice and centrifuged briefly at low speed before using.
   a. 10X RT Buffer: May cause eye, skin, and respiratory irritation; handle carefully.
   b. dNTP mix with dTTP (100 mM).
   c. RNase inhibitor (20 U/μL).
   d. Multiscribe™ RT enzyme (50 U/μL).
2. Nuclease-free water.
3. Stem-loop reverse transcription (RT) primers miRNAs of interest (Applied Biosystems, Foster City, CA).
4. TaqMan 2X Universal PCR Master Mix, No AmpErase UNG (Applied Biosystems, Foster City, CA).
5. qRT Primers (Applied Biosystems, Foster City, CA).

3. Methods

To investigate the change of miRNA expression profiles during ES cell differentiation into a tissue-specific cell type; it is important to maintain ES cells in an undifferentiated state prior to incubation under differentiation-promoting conditions. We have employed a standard method for the induction of pluripotent mouse ES D3 cells into neuronal cells. We employ TaqMan-based qRT-PCR to measure the expression of miRNAs in both differentiated and undifferentiated cells.

3.1. Cell Culture and Differentiation

1. Undifferentiated, pluripotent mouse ES D3 cells are cultured in gelatinized 25-cm flasks in a medium consisting of Knock-out DMEM, 15% fetal bovine serum – ES grade, 100 U/mL penicillin, 100 μg/mL streptomycin, 2 mM L-glutamine, 0.1 mM nonessential amino acids, 1400 U/mL murine leukemia inhibitory factor, and 55 μM 2-mercaptoethanol (*see* **Note 1**).
2. To induce differentiation, the medium is changed to differentiation medium (Iscove's Modified Dulbecco's Medium, 15% FBS, L-glutamine (2 mM), nonessential amino acids (0.1 mM), 100 units/mL penicillin, 100 μg/mL streptomycin) when the cells begin to form embryoid bodies (EBs) after 1–2 days of culture (*see* **Note 2**).
3. The medium is refreshed every 2 days. At that time, transfer cells into a 15-mL tube and let them settle to the bottom of

the tube for 10 min. Aspirate medium, add fresh medium, and carefully return the cells into the dish.

4. At day five, all-trans retinoic acid (Sigma, St. Lois, MO) (1×10^{-6}M) is added to the culture. The medium is changed and RA treatment is repeated on day seven.
5. At day 9, EBs are collected and seeded on poly-l-ornithine/fibronectin-coated flask in differentiation/Neurobasal plus B27 supplement 1:1 medium.
6. The next day, the medium is replaced with Neurobasal medium plus B27 supplement.
7. The cells are cultured for another 3 days.
8. In order to harvest the differentiated cells, the medium is aspirated, cells are washed with PBS, and 0.05% Trypsin/EDTA is added. Cells are incubated for 5 min at 37°C and 5% CO_2. The flask is gently taped to dislodge cells, and then 10 mL of differentiation medium is added to neutralize the trypsin.
9. Cells are transferred to a 15-mL tube and centrifuged for 5 min at 220 × g.
10. After centrifugation, cells are resuspended in 2 mL of Neurobasal medium and counted using hemacytometer.
11. The suspended cells are transferred into an RNase-free Eppendorf centrifuge tube, centrifuged, and medium is aspirated.
12. Cells are immediately frozen in liquid nitrogen (*see* **Note 3**).
13. Store cells in −80°C until RNA extraction.

3.2. MiRNA Isolation

Total RNAs are isolated from each cell sample using mirVana™ miRNA Isolation Kit (Ambion, Austin, TX) according to the manufacture's protocol.

1. Fresh or frozen cells (10^5–10^7) are washed by resuspension in ~1 mL PBS and repelleting. Place the washed cells on ice (*see* **Note 4**).
2. Remove the PBS wash or the RNAlater (if cells are stored in RNAlater).
3. Add 500 μL lysis/binding solution. Cells lyse immediately upon exposure to the lysis/binding solution.
4. Vortex or pipet vigorously for 5–10 s to completely lyse the cells and obtain a homogenous lysate.
5. Add 60 μL (1/10 the volume of lysis/binding buffer) miRNA homogenate additive to the homogenate and mix well by vortexing for 5–10 s or inverting the tube several times.

6. Keep the homogenate on ice for 10 min.
7. Add 600 μL acid-phenol/chloroform to each tube. The volume is equal to the lysis/binding buffer before miRNA homogenate additive addition (*see* **Note 5**).
8. Mix thoroughly by inverting or vortex the mixtures for approximately 30–60 s.
9. Centrifuge the tube at 10,000 × g at room temperature for 5 min to separate the aqueous phase from the organic phase. If the interface between the aqueous and organic phases is not compact after the centrifugation, a second round of centrifugation at the same speed and temperature should be performed to form a sharp interface.
10. Carefully remove the upper aqueous phase, being careful not to disturb the lower organic phase, and transfer to a new 1.5-mL tube. Write down the total volume of the upper aqueous phase transferred to the new samples.
11. Preheat elution solution or nuclease-free water to 95°C for later use in eluting the RNA from the filter at the end of the procedure.
12. Add 1.25 volumes of the aqueous phase of room temperature 100% ethanol to the aqueous phase. For example, if a total of 500 μL aqueous phase is recovered from step 10, then 575 μL ethanol should be added.
13. Mix well by vortexing or inverting several times.
14. For each sample, place a filter cartridge into a new collection tube provided with the kit. Pipet the lysate/ethanol solution onto the filter cartridge. A maximum of 700 μL of the lysate/ethanol solution can be loaded into the filter cartridge at any one time. If you have more than 700 μL of the solution, you can repeat the additions followed by centrifugation multiple times.
15. Centrifuge at 10,000 × g for about 15 s. Discard the flow-through and place the filter cartridge back into the same tube. Repeat this procedure until all of the lysate/ethanol solution has passed through the filter.
16. Use washing solution 1 to wash the filter cartridge. Apply 700 μL of miRNA washing solution 1 to the filter cartridge and centrifuge for approximately 5–10 s. Dispense of the flow-through and place the filter cartridge back into the same tube.
17. Apply 500 μL of miRNA wash solution 2/3 and centrifuge the solution through the filter cartridge as detailed in the previous step.
18. Repeat step 17 with a second aliquot with a volume of miRNA wash solution 2/3 used in step 17.

19. After discarding the flow-through from the previous step, put the filter cartridge back into the same collection tube and centrifuge the assembly for 1 min at 10,000 × g at room temperature. This removes residual fluid from the filter.
20. Transfer the filter cartridge to a newly marked collection tube.
21. Apply 100 μL of preheated 95°C elution solution or nuclease-free water to the center of the filter.
22. Incubate at room temperature for 30 s to 1 min.
23. Centrifuge the tube for 20–30 s at 10,000 × g to elute the total RNAs from the filter.
24. Remove the filter cartridge and mix the recovered RNAs by gently flicking the tube.
25. Briefly centrifuge again to collect the entire RNA-containing solution at the bottom of the tube.
26. Use a NanoDrop ND-1000 (NanoDrop Technologies, Wilmington, DE) to determine the quality and quantity of the total RNAs (*see* **Note 6**).
27. RNA samples are stored in a −80°C freezer until qRT-PCR analysis.

3.3. Quantitative Real-Time PCR (qRT-PCR) Analysis

A two-step TaqMan-based real-time PCR quantification is employed to determine and monitor miRNA expression during mouse ES cell differentiation. In the first step, an miRNA-specific stem-loop primer will be used to reverse transcribe mature miRNA to a cDNA sequence using Multiscribe^{TM} reverse transcriptase. A standard real-time PCR will be employed to perform the second step for determining miRNA expression using a forward miRNA-specific primer, a universal reversed primer, and FAM dye-labeled TaqMan probes (31). During this protocol, mouse small RNA snoRNA 135 serves as a reference gene for calculating the relative expression levels of each targeted miRNA.

1. Take the TaqMan microRNA Reverse Transcription Kit reagents and Reverse Transcription Primers (RT primers) out of the −20°C freezer.
2. Allow the kits and primers to thaw on ice. After thawing, briefly centrifuge to collect the reagents and primers in the bottom of the tubes.
3. In a PCR tube (0.2-mL tube), add the following amount of reagents into one reaction for preparing a RT master mix: 4.16 μL nuclease-free water, 0.19 μL RNase inhibitor, 1.5 μL 10 × RT buffer, 0.15 μL dNTP mix (100 mM), and 1.00 μL reverse transcriptase enzyme.

4. Gently mix the reagents by flicking the tube and briefly centrifuge.
5. Place the RT master mix back on ice while preparing the miRNA reaction.
6. Add 100–500 ng of total RNAs into each RT master mix and then add RNase-free water to a total of 14 μL.
7. Add 1 μL of RT primers to the appropriate tube bringing the total volume per tube to 15 μL. Gently mix the tube by flicking and centrifuge briefly.
8. Incubate for 5 min on ice or until ready to load the thermal cycler.
9. The reaction is incubated at 16°C for 30 min followed by 42°C for 30 min and 85°C for 5 min. Finally, the reverse transcription reaction holds at 4°C.
10. After completion of the reverse transcription, add 80 μL of nuclease-free water to the RT-PCR products (*see* **Note 7**).
11. Prepare a master mix in a new 0.2-mL PCR tube for real-time PCR. In a new PCR tube, add the following components to make a total of 20 μL volume reaction: 6 μL of nuclease-free water, 10 μL of 2 × PCR mixture, 2 μL of RT-PCR products (after addition of water), and 2 μL RT Primer. Three replicates need to be run for each sample.
12. Load the PCR solution into a 96-well PCR plate.
13. Briefly centrifuge the plate to collect the PCR solution in the bottom of the sample well in the plate
14. The reactions are incubated in a 96-well plate at 95°C for 10 min, followed by 40 cycles of 95°C for 15 s and 60°C for 60 s. This should take approximately 2 h.
15. After reactions have been completed, the threshold is manually set and the threshold cycle (C_T) is automatically recorded. The C_T is defined as the fractional cycle number at which the fluorescence signal passes the fixed threshold (31). All reactions are conducted in triplicate.
16. Based on the qRT-PCR results, the relative miRNA expression data are analyzed using the $\Delta\Delta C_T$ method and the differentially expressed miRNAs are identified.

4. Notes

1. To gelatinize flasks, add 0.1% gelatin to a 25-cm^2 Falcon Tissue Culture Flask (Fisher, Pittsburg, PA, Cat. No. 353014) and incubate at room temperature for 15 min.

After treatment, aspirate the gelatin solution and air-dry treated flasks for 30 min in the hood.

2. For differentiation of ES cells to neuronal cells, use BD Falcon Standard Dishes (Fisher, Pittsburg, PA, Cat No. 08-757-100D). Cells usually do not adhere to the bottom of the dish. Seed approximately 500,000 ES cells per 10 mL of differentiation medium to achieve optimal amount of embryoid bodies. Resuspend cells carefully but thoroughly to generate single-cell suspensions and minimize clumps.
3. Using fresh cells will produce better results. If it is necessary to store cells prior to the preparation of RNA, cells can be stored in RNAlater, or they can be pelleted and immediately frozen in liquid nitrogen. Then, these cells can be stored at −70°C or colder.
4. Cells always need to be incubated on ice to inhibit RNase and prevent RNA degradation.
5. Make sure to withdraw Acid-Phenol:Chloroform from the lower phase of the bottle because the upper phase consists of an aqueous buffer. The erroneous use of the upper phase will likely generate very poor RNA yield and quality.
6. The mirVana^{TM} miRNA Isolation Kit can be used to extract both total RNAs or small RNAs. Here, the protocol presented is for isolating total RNAs because total RNAs is of sufficient quality to be used for qRT-PCR analysis. If you need to isolate and enrich small RNAs, please refer to the manufacture's protocol.
7. After RT-PCR, the reverse transcription products must be diluted by 5–10 folds to avoid the potential interference from the high concentration of the stem-loop primer.

Acknowledgments

This work was partially supported by East Carolina University New Faculty Research Startup Funds Program (to BZ and XP) and a Science and Engineering Grant from DuPont (to BZ).

References

1. Ambros, V. (2001) microRNAs: Tiny regulators with great potential. *Cell* **107**, 823–826.
2. Bartel, D. P. (2004) microRNAs: Genomics, biogenesis, mechanism, and function. *Cell* **116**, 281–297.
3. Zhang, B. H., Pan, X. P., Cobb, G. P., and Anderson, T. A. (2006) Plant microRNA: A small regulatory molecule with big impact. *Dev Biol* **289**, 3–16.
4. Griffiths-Jones, S., Saini, H. K., van Dongen, S., and Enright, A. J. (2008) miRBase: Tools

for microRNA genomics. *Nucleic Acid Res* **36**, D154–D158.
5. Lewis, B. P., Burge, C. B., and Bartel, D. P. (2005) Conserved seed pairing, often flanked by adenosines, indicates that thousands of human genes are microRNA targets. *Cell* **120**, 15–20.
6. Xie, X. H., Lu, J., Kulbokas, E. J., Golub, T. R., Mootha, V., Lindblad-Toh, K., Lander, E. S., and Kellis, M. (2005) Systematic discovery of regulatory motifs in human promoters and 3′ UTRs by comparison of several mammals. *Nature* **434**, 338–345.
7. Zhang, B. H., Wang, Q. L., and Pan, X. P. (2007) microRNAs and their regulatory roles in animals and plants. *J Cell Physiol* **210**, 279–289.
8. Ambros, V. (2004) The functions of animal microRNAs. *Nature* **431**, 350–355.
9. Williams, A. E. (2008) Functional aspects of animal microRNAs. *Cell. Mol Life Sci* **65**, 545–562.
10. Zhang, B. H., Pan, X. P., and Anderson, T. A. (2006) microRNA: A new player in stem cells. *J Cell Physiol* **209**, 266–269.
11. Wang, Y. L., Keys, D. N., Au-Young, J. K., and Chen, C. F. (2009) microRNAs in embryonic stem cells. *J Cell Physiol* **218**, 251–255.
12. Li, Q. T., and Gregory, R. I. (2008) microRNA regulation of stem cell fate. *Cell Stem Cell* **2**, 195–196.
13. Hatfield, S., and Ruohola-Baker, H. (2008) microRNA and stem cell function. *Cell Tissue Res* **331**, 57–66.
14. Houbaviy, H. B., Murray, M. F., and Sharp, P. A. (2003) Embryonic stem cell-specific microRNAs. *Devl Cell* **5**, 351–358.
15. Forstemann, K., Tomari, Y., Du, T. T., Vagin, V. V., Denli, A. M., Bratu, D. P., Klattenhoff, C., Theurkauf, W. E., and Zamore, P. D. (2005) Normal microRNA maturation and germ-line stem cell maintenance requires loquacious, a double-stranded RNA- binding domain protein. *Plos Biology* **3**, 1187–1201.
16. Gangaraju, V. K., and Lin, H. F. (2009) microRNAs: Key regulators of stem cells. *Nat Rev Mol Cell Biol* **10**, 116–125.
17. Greco, S. J., and Rameshwar, P. (2007) microRNAs regulate synthesis of the neurotransmitter substance P in human mesenchymal stem cell-derived neuronal cells. *Proc Nat Acad Sci USA* **104**, 15484–15489.
18. Hammond, S. M., and Sharpless, N. E. (2008) HMGA2, microRNAs, and stem cell aging. *Cell* **135**, 1013–1016.
19. Hatfield, S. D., Shcherbata, H. R., Fischer, K. A., Nakahara, K., Carthew, R. W., and Ruohola-Baker, H. (2005) Stem cell division is regulated by the microRNA pathway. *Nature* **435**, 974–978.
20. Ivey, K. N., Muth, A., Amold, J., King, F. W., Yeh, R. F., Fish, J. E., Hsiao, E. C., Schwartz, R. J., Conklin, B. R., Bernstein, H. S., and Srivastava, D. (2008) MicroRNA regulation of cell lineages in mouse and human embryonic stem cells. *Cell Stem Cell* **2**, 219–229.
21. Kanellopoulou, C., Muljo, S. A., Kung, A. L., Ganesan, S., Drapkin, R., Jenuwein, T., Livingston, D. M., and Rajewsky, K. (2005) Dicer-deficient mouse embryonic stem cells are defective in differentiation and centromeric silencing. *Genes Dev* **19**, 489–501.
22. Krichevsky, A. M., Sonntag, K. C., Isacson, O., and Kosik, K. S. (2006) Specific microRNAs modulate embryonic stem cell-derived neurogenesis. *Stem Cells* **24**, 857–864.
23. Kuwabara, T., Hsieh, J., Nakashima, K., Taira, K., and Gage, F. H. (2004) A small modulatory dsRNA specifies the fate of adult neural stem cells. *Cell* **116**, 779–793.
24. Lakshmipathy, U., Love, B., Goff, L. A., Jornsten, R., Graichen, R., Hart, R. P., and Chesnut, J. D. (2007) microRNA expression pattern of undifferentiated and differentiated human embryonic stem cells. *Stem Cells Dev* **16**, 1003–1016.
25. Murchison, E. P., Partridge, J. F., Tam, O. H., Cheloufi, S., and Hannon, G. J. (2005) Characterization of dicer-deficient murine embryonic stem cells. *Proc Nat Acad Sci USA* **102**, 12135–12140.
26. Park, J. K., Liu, X., Strauss, T. J., McKearin, D. M., and Liu, Q. H. (2007) The miRNA pathway intrinsically controls self-renewal of Drosophila germline stem cells. *Curr Biol* **17**, 533–538.
27. Suh, M. R., Lee, Y., Kim, J. Y., Kim, S. K., Moon, S. H., Lee, J. Y., Cha, K. Y., Chung, H. M., Yoon, H. S., Moon, S. Y., Kim, V. N., and Kim, K. S. (2004) Human embryonic stem cells express a unique set of microRNAs. *Dev Biol* **270**, 488–498.
28. Tang, F. C., Hajkova, P., Barton, S. C., Lao, K. Q., and Surani, M. A. (2006) microRNA expression profiling of single whole embryonic stem cells. *Nucleic Acids Res* **34**, e9.
29. Wang, Y., Baskerville, S., Shenoy, A., Babiarz, J. E., Baehner, L., and Blelloch, R. (2008) Embryonic stem cell-specific microRNAs regulate the G1-S transition and promote rapid proliferation. *Nat Genet* **40**, 1478–1483.

30. Zou, G. M., and Yoder, M. C. (2005) Application of RNA interference to study stem cell function: current status and future perspectives. *Biol Cell* 97, 211–219.
31. Chen, C. F., Ridzon, D. A., Broomer, A. J., Zhou, Z. H., Lee, D. H., Nguyen, J. T., Barbisin, M., Xu, N. L., Mahuvakar, V. R., Andersen, M. R., Lao, K. Q., Livak, K. J., and Guegler, K. J. (2005) Real-time quantification of microRNAs by stem-loop RT-PCR. *Nucleic Acids Res* **33**, e179.

Chapter 17

Engineering Human Mesenchymal Stem Cells to Release Adenosine Using miRNA Technology

Gaoying Ren and Detlev Boison

Abstract

Adenosine is an important modulator of metabolic activity with powerful tissue- and cell-protective functions. Adenosine kinase (ADK), the major adenosine-regulating enzyme, is critical to adapt its intra- and extra-cellular levels in response to environmental changes. Lentiviral RNAi-mediated down-regulation of ADK in human mesenchymal stem cells (hMSCs) has therefore been considered an effective tool for engineering therapeutically effective adenosine-releasing cell grafts that could constitute patient-identical autologous implants for clinical application. We constructed lentiviral vectors that coexpress miRNA directed against ADK and an emerald green fluorescent protein (EmGFP) reporter gene. Following lentiviral transduction of hMSCs, we demonstrated up to 80% down-regulation of ADK and 98% transduction efficiency. Transduced hMSCs continued to express EmGFP after 4–6 consecutive passages and EmGFP-positive hMSC grafts survived in the hippocampal fissure of mouse brains and provided efficient adenosine-dependent neuroprotection in a mouse model of seizure-induced cell loss.

Key words: Adenosine, adenosine kinase, epilepsy, kainic acid, RNAi, lentivirus, human mesenchymal stem cells, cell therapy.

1. Introduction

Adenosine is an endogenous neuromodulator of the brain with potent inhibitory and neuroprotective properties (1). Adenosine levels are largely regulated by adenosine kinase (ADK) (2), and increases in ADK lead to increased seizure activity (3, 4), while reduced levels of ADK augment endogenous adenosine and reduce seizure activity (5).

Human mesenchymal stem cells (hMSCs) are an excellent candidate for patient-derived autologous cell transplantation

B. Zhang, E.J. Stellwag (eds.), *RNAi and MicroRNA-Mediated Gene Regulation in Stem Cells*, Methods in Molecular Biology 650,
DOI 10.1007/978-1-60761-769-3_17,

because of their differentiation potential and immunocompatibility (6, 7). Using lentiviral expression vectors, gene expression can effectively be achieved in hMSCs (8), and more importantly, stable gene expression can be continued throughout differentiation (9). Therefore, we engineered hMSCs to release adenosine using a lentivirus expressing miRNA directed against the major adenosine-removing enzyme ADK. This protocol will benefit the future engineering of patient-identical hMSCs for the therapeutic delivery of adenosine for autologous cell-transplantation approaches.

We have constructed lentiviral vectors, which express anti-ADK miRNA, an emerald green fluorescent protein (EmGFP) reporter gene, and a blasticidin-resistance gene for stable clone selection. After transducing hMSCs with lentiviral construct and further blasticidin selection, the knockdown of ADK is quantitatively analyzed by immunoblotting and Western blot analysis. Immunofluorescence detection allows the evaluation of the transduction efficiency of hMSCs and EmGFP-expressing hMSC grafts in mouse brain after transplantation (10).

2. Materials

2.1. Lentiviral Construction for the Expression of Anti-ADK miRNA

1. pcDNA™ 6.2-GW/EmGFP-miR expression vector.
2. pDONR™ 221 vector.
3. pLenti6/V5-DEST destination vector.
4. BP Clonase™ II enzyme mix, LR Clonase™ II enzyme mix. All from Invitrogen (Carlsbad, CA).

2.2. Lentiviral Production, Titer, and Transduction

2.2.1. Lentiviral Production

1. Viralpower™ Packaging Mix (Invitrogen, Carlsbad, CA).
2. Lipofectamine™ 2000 (Invitrogen, Carlsbad, CA), store at 4°C.
3. Geneticin® (Invitrogen, Carlsbad, CA): 50 mg/mL, 100x stock, store at −20°C.
4. Opti-MEM® I Reduced Serum Medium (Gibco/BRL, Bethesda, MD).
5. Dulbecco's phosphate-buffered saline (PBS, Gibco/BRL, Bethesda, MD).
6. 293FT Cells (Invitrogen, Carlsbad, CA).
7. 293FT culture medium: Dulbecco's Modified Eagle's Medium (DMEM) supplemented with 10% fetal bovine serum (FBS), 0.1 mM MEM nonessential amino acids, 2 mM of L-glutamine, 1% penicillin and streptomycin (all from Gibco/BRL, Bethesda, MD), and 500 μg/mL geneticin® (*see* **Note 1**).

2.2.2. Lentiviral Titer

1. Polybrene® (Invitrogen, Carlsbad, CA): 6 mg/mL stock, aliquot and store at −20°C.
2. Blasticidin (Invitrogen, Carlsbad, CA): 10 mg/mL stock, aliquot and store at −20°C.
3. HT1080 cells (ATCC, Manassas. VA).
4. HT1080 culture medium: Modified Eagle's Minimal Essential Medium supplemented with 1 mM sodium pyruvate, 0.1 mM MEM nonessential amino acids, and 10% nonheat inactivated FBS (ATCC, Manassas. VA).
5. HT1080 selection medium: HT1080 culture medium contains 10 ug/mL of blasticidin.
6. FACScan (BD Bioscience, San Jose, CA).
7. Cresyl violet: Prepare 1% cresyal violet solution in 10% ethanol (Sigma).

2.2.3. Lentiviral Transduction

1. Human bone marrow-derived mesenchymal stem cells (hMSCs, Cambrex, Walkersville, MD).
2. hMSC culture medium: DMEM supplemented with 10% FBS for human mesenchymal cells (StemCell Technologies, Vancouver, B.C. Canada), 2 mM L-glutamine, 0.1 mM MEM nonessential amino acids solution, and 1 mM sodium pyruvate.
3. hMSC selection medium: hMSC culture medium contains 5 μg/mL of blasticidin.
4. Phase contrast fluorescent microscope (Nikon, ELLIPSE, TS100, Japan).

2.3. Quantitative Analysis of EmGFP Expression Efficiency

1. Paraformaldehyde (Fisher): Prepare 4% (w/v) solution in PBS.
2. Poly-L-lysine solution (Sigma).
3. 8-well chamber slides (Nunc, Rochester, NY).
4. Fluorescence mounting medium containing DAPI (Vector Laboratories, Inc. Burlingame, CA).
5. Leica DMLB light and fluorescent microscope (Zeiss Jena, Germany).
6. High-resolution Zeiss AxioCam camera (Zeiss Jena, Germany).

2.4. Preparation of Samples for SDS-Electrophoresis

1. Cell lysis buffer: 50 mM Tris-HCL, pH 8.0, 150 mM NaCl, and 1% Nonidet P-40. Store at 4°C.
2. Protease inhibitor cocktail (Sigma, St. Louis, MO), aliquot and store at −20°C.
3. Bradford reagent for protein quantitation (Sigma), store at 4°C.

4. Laemmli Sample Buffer 2X (Bio-Rad, Hercules, CA): Add 50 μL of β-mercaptoethanol in 950 μL of Laemmli Sample Buffer before use.

2.5. SDS-Electrophoresis

1. 12.5% Criterion Tris-HCl Gels (Bio-Rad, Hercules, CA), store at 4°C (*see* **Note 2**).
2. Precision Plus Protein Dual Color Standards (Bio-Rad, Hercules, CA). Aliquot and store at −20°C.
3. Electrophoresis buffer (10X): 0.25 M Tris base, 1.92 M Glycine, and 1% SDS. Store at room temperature.
4. Criterion cell for electrophoresis (Cell/Plate Blotter System, Bio-Rd, Hercules, CA).

2.6. ADK Knockdown Quantification by Western Blot

1. Criterion Blotter (Criterion Cell/Plate Blotter System, Bio-Rd, Hercules, CA).
2. Transfer buffer (10X): 0.25 M Tris base, 1.92 M Glycine. Store at 4°C. 1000 mL 1X transfer buffer: 100 mL 10X transfer buffer, 200 mL methanol, and 800 mL H_2O. Chill to 4°C before use.
3. Immuno Blot PVDF Membrane (Bio-Rad, Hercules, CA).
4. Phosphate-buffered saline (PBS): Prepare 10x stock with 1.37 M NaCL, 27 mM of KCL, 100 mM Na_2HPO_4, and 18 mM KH_2PO (adjust to pH 7.4 with HCL if necessary) and autoclave before storage at room temperature. Prepare working solution (1 × PBS) by dilution of one part with nine parts water.
5. Phosphate-buffered saline with 0.05% Triton^{R}X-100 (PBST): Add 2.5 mL of 20% Triton^{R}X-100 (Sigma) to 1000 mL 1XPBS.
6. Rabbit anti-ADK primary antibody (Detlev Boison's lab, Portland, OR), aliquot and store at −20°C.
7. Blocking buffer: 1% (w/v) BSA (Sigma) in PBS. Store at −20°C.
8. Secondary antibody: Antirabbit IgG-conjugated to horse radish peroxidase (Cell Signaling Technology, Inc., Danvers, MA). Store at −20°C.
9. Rabbit polyclonal beta actin antibody (Loading Control, Abcam, Cambridge, MA). Aliquot and store at −20°C.

2.7. Stripping and Reprobing Blots with β-Actin for ADK Quantification

1. Stripping buffer: 200 mM NaOH. Store at room temperature.
2. Enhanced chemiluminescent (ECL) reagents from NEN, Life Sciences, Boston, MA). Store at 4°C.
3. Kodak Image Analysis Software (Eastman Kodak Company, NY).

2.8. Cell Transplantation

1. Cyclosporine A: Prepare working solution of 2.5 mg/mL in 0.9% NaCL. Store at 4°C in dark.
2. Kopf stereotactic frame (KOPF Instrument, Wood Dale, Illinois).
3. Glass capillary (inner diameter of tip: 70–90 μm, Humagen IVF, Charlottesville, VA).

2.9. Graft Histology

1. 2-Methylbutane (Sigma).
2. Formaldehyde solution, 10% w/w (Fisher).

3. Methods

3.1. Lentiviral Construction for the Expression of Anti-ADK miRNA

1. Design several different pre-miRNA sequences with homology to your target cDNA according to published recommendations (11, 12). In addition, a randomized, scrambled control (SC) sequence is synthesized (Invitrogen, Carlsbad, CA). An example of the designed sequences for ADK is shown in **Fig. 17.1A**.
2. Clone the selected sequences into pcDNA^{TM} 6.2-GW/EmGFP-miR expression vectors, which contain an emerald green fluorescence protein (EmGFP) reporter gene and a blasticidin-resistance gene, thus allowing coexpression of the respective miRNA with EmGFP and selection of stably transduced cells with blasticidin (*see* **Fig. 17.1B**). Sequence all constructs to confirm their structure.

A

p236 (human/mouse/rat)	5'-AATGTCTTTAGTCTCAAAGCC-3'
p239 (human)	5'-AAATGTTGCTGCTTTGTGTGG-3'
p240 (human)	5'-AAACGGTGCAGATAGATTCAA-3'
p241 (human)	5'-TTCTCAGGAAGCAATTAGGGC-3'
p242 (human)	5'-ACAAGCATTACCATACCCTTG-3'

B

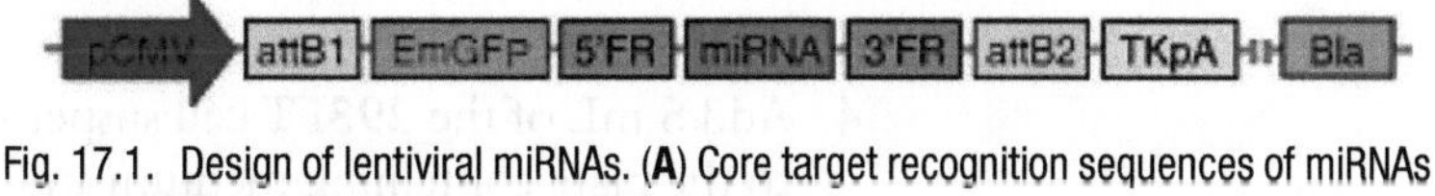

Fig. 17.1. Design of lentiviral miRNAs. (**A**) Core target recognition sequences of miRNAs from human only (p239–p242), or human, mouse, and rat (p236) ADK cDNA. (**B**) Schematic gene map of the pcDNA^{TM} 6.2-GW/EmGFP-miR expression vectors into which the miRNA sequences derived from (**A**) were cloned. pCMV: cytomegalovirus promoter, attB1: first attB substrate, EmGFP: emerald green fluorescence protein, 5′FR: 5′ flanking region, miRNA, 3′ FR: 3′ flanking region, attB2: second attB substrate, TKpA: thymidine kinase polyadenylation signal, Bla: blasticidin-resistance gene. (Reproduced from (10) with permission from ELSEVIER Science).

3. Transfer the pre-miRNA expression cassettes from the pcDNA™ 6.2-GW/EmGFP-miR expression clones into a pLenti6/V5-DEST destination vector for the production of a lentiviral expression clone:
   a. Perform a recombination reaction between the *att*B substrate in the pcDNA™ 6.2-GW/EmGFP-miR expression clone (*see* **Fig. 17.1B**) and the *att*P substrate in a pDONR™ 221 vector using BP Clonase-II™ to generate an entry clone.
   b. Perform another recombination reaction between the resulting entry clone containing the *att*L substrate and the pLenti6/V5-DEST vector containing the *att*R substrate.

3.2. Lentiviral Production, Titer, and Transduction

3.2.1. Lentiviral Production

Thaw and plate 293FT cells into a 10-cm dish with culture medium without Geneticin®. After 4 h, replace the medium with culture medium containing 500 μg/mL Geneticin®. Cells are maintained and passaged at a ratio of 1:5 for at least 3 passages (*see* **Note 3**).

1. For each transfection sample, prepare DNA–Lipofectamine™ 2000 complexes as follows:
   a. In a 5-mL sterile tube, dilute 9 μg ViraPower™ Packaging Mix and 3 μg plenti6/V5-GW/miR expression plasmid DNA (total 12 μg) in 1.5 mL of Opti-MEM® I Reduced Serum Medium. Mix gently.
   b. In a separate sterile 5-mL tube, mix Lipofectamine™ 2000 gently before use and then dilute 36 μL in 1.5 mL of Opti-MEM® Reduced Serum Medium. Mix gently and incubate for 5 min at room temperature. Then combine the diluted DNA with the diluted Lipofectamine™ 2000. Mix gently. Incubate for 20 min at room temperature to allow the DNA–lipid complexes to form.
2. While DNA–lipid complexes are forming, trypsinize and count the 293FT cells. Resuspend the cells at a density of 1.2×10^6 cells/mL in culture medium without antibiotics (*see* **Note 4**).
3. Add DNA–lipid complexes to a 10-cm tissue culture plate containing 5 mL of culture medium. (Do not include antibiotics in the medium).
4. Add 5 mL of the 293FT cell suspension (6×10^6 total cells) to the plate containing medium and DNA–Lipofectamine™ 2000 complexes and mix gently by rocking the plate back and forth. Incubate the cells overnight at 37°C in a CO_2 incubator.
5. After 8 h or the next day, replace the medium with fresh culture medium very gently (*see* **Notes 5** and **6**).

6. Harvest virus-containing supernatant 48–72 h posttransfection by removing medium to a 15-mL sterile, capped, conical tube.
7. Centrifuge at 500×*g* for 5 min at 4°C. Perform filtration step, if desired.
8. Pipet viral supernatant into 1.5 mL eppendorf tubes to make 1 mL aliquot. Store viral stocks at −80°C. If higher titer is desired, the virus supernatant can be concentrated first, then aliquoted, and stored at −80°C.

3.2.2. Lentiviral Titer

1. The day before transduction, (Day 1), trypsinize and count the HT1080 cells and plate them (2 × 10^5/well) in a 6-well plate such that they will be 30–50% confluent at the time of transduction. Incubate cells at 37°C overnight.
2. On the day of transduction (Day 2), thaw lentiviral stocks and prepare 10-fold serial dilutions ranging from 10^{-2} to 10^{-6} (*see* **Note 7**). For each dilution, dilute the lentiviral stocks into complete culture medium containing 6 μg/mL of polybrene® to a final volume of 1 mL. Do not vortex.
3. Remove the culture medium from the cells. Mix each dilution gently by inversion and add to one well of cells (total volume=1 mL). Incubate at 37°C overnight.
4. The following day (Day 3), remove the medium containing the virus and replace with 2 mL of complete culture medium.
5. The following day (Day 4), process to Steps 6–7 for EmGFP titering method or proceed to Steps 8–12 for blasticidin titering method.
6. Determine the titer by flow cytometry on Day 4 for titering EmGFP. For each viral dilution well of the 6-well plate, trypsinize and wash the cells with PBS and resuspend the cells in PBS with 1% FBS at a concentration of 10–,500 cells/μL (*see* **Note 8**).
7. Using a flow cytometry system, determine the percentage of GFP-positive cells for each dilution. Titer is expressed as transducing units (TU)/mL. Use the formula of $(F \times C/V) \times D$ to calculate the titer: F = frequency of GFP-positive cells, C = total number of cells in the well at the time of transduction, V = volume of inoculum in mL, D = lentiviral dilution.
8. Using blasticidin selection, remove the medium on Day 4 and replace HT1080 selection medium to select stably transduced cells. Replace selection medium every 3–4 days.
9. After 10–12 days of selection (Days 14–16), you should see only dead cells in the mock well and discrete blasticidin-

resistant colonies in one or more of these dilution wells. Remove the medium and wash the cells twice with PBS.

10. Add 1 mL of a 10% Cresyl violet solution into the well and incubate for 10 min at room temperature.
11. Remove the Cresyl violet stain and wash the cells with water.
12. Count the blue-stained colonies and determine the lentiviral stock titer. Lentiviral titer (TU/mL)=numbers of colonies × lentiviral dilution.

3.2.3. Lentiviral Transduction

1. hMSCs at early passages (passages 1 or 2, *see* **Note 9**) are cultured at 37°C, 5% CO_2, and 95% humidity in complete growth medium and cells are passaged when reaching 90% confluence (5–7 days) at a ratio of 1:3.
2. One day before transduction, hMSCs are trypsinized and plated in a 6-well plate at a density of 2×10^5 cells/well.
3. The next day (Day 1), medium is removed and replaced with 1 mL of hMSC culture medium containing lentivirus at an MOI of 5 (*see* **Note 10**) and 6 μg/mL polybrene®.
4. The next day (Day 2), medium is replaced with fresh hMSC culture medium.
5. On Day 4, medium is replaced with hMSC selection medium containing blasticidin to select stably transduced cells (*see* **Note 11**).
6. Change fresh hMSC selection medium every 3–4 days and monitor EmGFP fluorescence of transduced hMSCs daily using a phase-contrast fluorescent microscope. The selection will be done 7–10 days after selection. The cells are either maintained and passaged or frozen in a vial for future use at 7–10 days after Blasticidin selection.

3.3. Quantitative Analysis of EmGFP Expression Efficiency

1. Before and after selection, hMSCs are trypsinized and plated in duplicates on poly-L-lysine-coated 8-well chamber slides at a density of 4×10^4 cells per well.
2. At 24 h, the cells are fixed with 4% paraformaldehyde solution in PBS, followed by 2 times PBS wash.
3. The nucleus of the cells is labeled by mounting the chamber slide with fluorescence mounting medium containing DAPI.
4. For quantitative analysis of EmGFP expression efficiencies, four areas of each well are randomly selected and the number of total cells (DAPI) and EmGFP-expressing cells is observed with a Leica DMLB microscope, and photographed using a high-resolution Zeiss Axiocam camera. An example of the results is shown in **Fig. 17.2**.

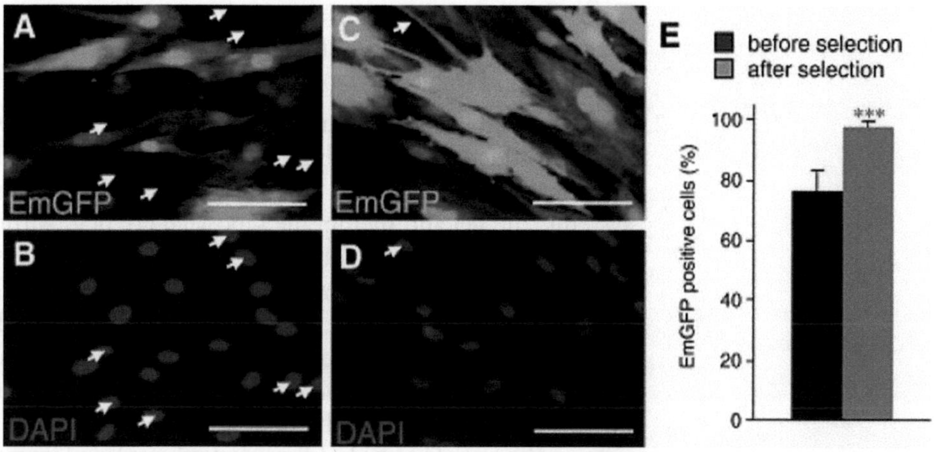

Fig. 17.2. Lentiviral transduction efficiency in hMSCs. EmGFP expression (*green*) in hMSCs was assessed 5 days after transduction with vector H239 (**A**) or 7 days after selecting these cells with blasticidin (5 μg/mL) (c). Cells were counter-stained with DAPI to visualize cell nuclei (**B, D**). Arrows indicate EmGFP-negative cells. (**E**) Before blasticidin selection, 76 ± 7.3% of the cells were positive for EmGFP, while after selection the percentage of EmGFP-positive cells had increased to 98 ± 2.0%. $P < 0.001$, chi-square test, $n = 8$. (Reproduced from (10) with permission from ELSEVIER Science).

3.4. Preparation of Samples for SDS-Electrophoresis

1. Seven days after blasticidin selection, transduced hMSCs are trypsinized and harvested in a 1.5-mL tube.
2. Cells are centrifuged for 10 min at 500×*g*, the pellets washed with PBS twice, and lysed with 50 μL of chilled cell-lysis buffer containing 10 μL/mL of protease inhibitor cocktail; they are then incubated on ice for 30 min, during which they are vortexed 2–3 times, and then centrifuged for 30 min at 500×*g*.
3. The aqueous supernatants are collected and processed for protein quantitation using Bradford Reagent according to the instruction of the manufacturer (Sigma).

3.5. SDS-Electrophoresis

1. To make 50 μL loading samples containing 30 μg proteins, add 25 μL supernatant containing 30 μg of protein and 25 μL 2X loading buffer; mix and heat at 95°C for 5 min.
2. Add 1x running buffer into the upper and lower chambers of the gel unit and load 50 μL of each sample in one well. Include one well of prestained molecular weight markers.
3. Complete the assembly of the gel unit and connect to a power supply. Individual samples, each containing 30 μg protein, are separated on a precast SDS/12.5% polyacrylamide gel in a Tris/HCl buffer (pH 7.4) at 150 V for 90 min.

3.6. ADK Knockdown Quantification by Western Blot

1. After the gel has been electrophoresed, it is soaked in transfer buffer for 5–10 min. A PVDF membrane, which is cut the same size as the gel, is soaked in methanol for 1 min.

Two filter papers, which match the size of the gel, are soaked in the transfer buffer just before assembling the blotting sandwich.

2. In an assembly tray, about 700 mL of cold 1x transfer buffer is added. Two blotting pads are soaked in the buffer and air is removed using a roller. In tank (or wet) transfer systems, the gel and membrane sandwich are held within a gel holder cassette and submerged entirely under transfer buffer. To make the blot sandwich, a gel holder cassette is placed in the tray with the black side on the bottom. The presoaked blotting pad, filter paper, gel, PVDF membrane, and another filter paper and fiber pad are assembled. After removing all air bubbles, the cassette is closed with the red side on the top. The gel holder cassette is replaced immediately in the tank containing 1.3 mL of prechilled 1x transfer buffer. The black side of the cassette faces the black plate electrode, which is the anode, and the red side faces the red plate electrode, which is the cathode.
3. Electrophorese the transblot at 4°C (in a cold room) using an ice pack within the transfer buffer at 100 V for 45 min.
4. After the trans blot, the membrane is incubated in blocking buffer for 1 h at room temperature on a shaker.
5. The blot is incubated with polyclonal rabbit antiserum against ADK (1:4000) in blocking buffer containing 0.05% sodium azide at 4°C on a shaker, overnight (*see* **Note 12**).
6. The blot is washed with PBST, 5 × 20 min.
7. The blot is incubated for 1 h at room temperature with freshly prepared horseradish peroxidase (HRP)-linked antirabbit antibody diluted at 1:8000 with blocking buffer.
8. The blot is washed sequentially with PBST, 3 × 20 min, then high-salt TBST containing 500 mM NaCL for 20 min, and then with PBST again, two times for 20 min.
9. During the final wash, ECL reagents are warmed to room temperature, and upon finishing the final wash, 2 mL of each portion of ECL is mixed in a container which fits the blot size and the blot is added to it, which is then rotated by hand for 20 s to ensure even coverage.
10. In the dark room, the blot is removed from the ECL reagents and placed between the transparent plastic sheets that have been cut to the size of an X-ray film cassette.
11. The plastic sheet containing the blot is then placed in an X-ray film cassette with film for a suitable exposure time,

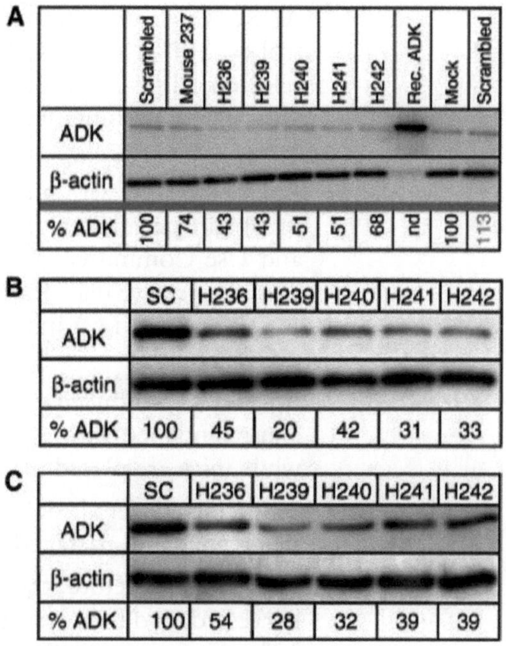

Fig. 17.3. ADK-knockdown in hMSCs by lentiviral RNAi. (a) In a primary assessment, Western blot analysis was performed on cell lysates (30 μg each) 72 h after lentiviral transduction with MOI's of 25. ADK staining top and β-actin staining bottom; Controls: Mock transduction, transduction with scrambled control virus, or with mouse-specific virus 237; rec. ADK = recombinant ADK. H236–H242: hMSCs transduced with 5 different human anti-ADK miRNA vectors. (**B, C**) Western blots from samples (30 μg, each) derived from duplicate transductions of hMSCs with a scrambled control vector (SC) or with the anti-ADK miRNA vectors H236–H242. Samples were prepared 7 days after blasticidin selection. The blots were probed with antibodies directed against ADK (*top*) or β-actin. Residual ADK (%ADK) was calculated from scanned intensities of bands in the Western blots, and normalized to β-actin and respective control hMSCs (= 100%). Note that blasticidin-selected hMSC cells, H239, displayed the strongest reduction of ADK expression. (Reproduced from (10) with permission from ELSEVIER Science).

typically, starting from 30 s to a few minutes. An example of the results produced is shown in **Fig. 17.3**.

3.7. Reprobing Blots with β-Actin for ADK Quantification

1. The blot is washed with H_2O for 5 min and then incubated in 200 mM NaOH for 5 min with gentle shaking, and washed with H_2O, 3 × 5 min.
2. After incubating with blotting buffer for 1 h, the blot is reprobed with a β-actin rabbit polyclonal antibody (1:2000) diluted in blocking buffer at 4°C, overnight.
3. Follow the Steps 6–11 in **Section 3.6**.
4. The intensities of the bands in each Western blot are quantified using the Kodak Image Analysis Software (Eastman Kodak Company, NY). Immunoreactivity to β-actin is used as an internal standard to calculate relative amounts of

ADK, which are then normalized to the scrambled control (= 100%) samples. The experiment for the blasticidin-selected cells is performed in duplicate. An example of ADK knockdown quantitation is shown in **Fig. 17.3**.

3.8. Cell Transplantation

1. Before starting with any animal experimentation you need to obtain permission from your Institutional Animal Care and Use Committee. Adult male C57BL/6 mice weighing 25–30 g receive daily immunosuppression with cyclosporine A (15 mg/kg, i.p.) initiated 2 days prior to cell transplantation.
2. Anesthesia is induced with 3% isoflurane, 67% N_2O, 30% O_2 and maintained with 1.5% isoflurane, 68.5% N_2O, 30% O_2, while mice are placed in a Kopf stereotactic frame.
3. Blasticidin-selected stable EmGFP-expressing hMSCs are maintained and passaged in blasticidin selection medium.
4. Immediately before transplantation, cells are trypsinized, harvested, and then resuspended at a concentration of 2.5×10^4 cells per μL in culture medium. Cell injections (5×10^4 cells per mouse) are performed using a glass capillary (inner diameter of tip: 70–90 μm). The cells are slowly injected in a volume of 2 μL using a drill hole above the left hippocampus and a single diagonal injection tract spanning from coordinate (AP + 1.6; ML + 1.2; DV 0.0) to coordinate (AP – 2.8; ML – 1.75; DV – 4.0), thus depositing the cells within the infrahippocampal cleft of the to-be-injured brain hemisphere (13). Cells are slowly injected (1 μL/min) while withdrawing (1 mm/min) the capillary. The capillary is fully retracted 5 min after injection to avoid reflux of cells.

3.9. Histological Analysis of Grafts

1. Brains are obtained by surgery 1 week after transplantation and immediately frozen in 2-methylbutane (– 30°C), and sectioned at 12 μm on a cryostat. The sections are stored at – 80°C until use.
2. Coronal sections – 1.7 to – 2.1 mm caudal to bregma (14) are air dried (15 min), postfixed in 10% formaldehyde (15 min), washed twice with PBS, and then processed for histological analysis.
3. To visualize the intrahippocampal implants, graft-based EmGFP fluorescence is determined after mounting the slides with mounting medium containing DAPI. Images are visualized using a Leica microscope DMLB under an Ex/Em wavelength of 500/550 nm (green) and photographed using a high-resolution Zeiss Axiocam camera. An example of the graft histology is shown in **Fig. 17.4**.

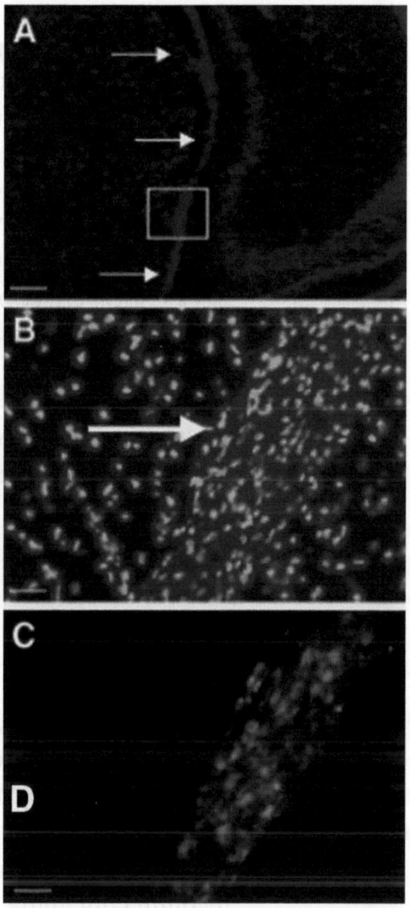

Fig. 17.4. Morphology of cell grafts. Representative DAPI-stained coronal brain sections from mice 8 days after transplantation of H239-transduced hMSCs cells. (**A**) Composite fluorescence image at lower magnification showing general graft morphology and location within the infrahippocampal cleft (*arrows*). (**B**) Fluorescence image of the boxed area in panel A showing the high density of nuclei from the graft (*arrow*). Note the presence of intact nuclei indicative of graft survival. (**C**) Same image as (**B**) viewed under a GFP-specific filter. (**D**) Different image selected at 30 days after cell transplantation viewed under a GAP-specific filter. Scale bars: A: 100 μm, B, C, D: 25 μm. (Reproduced from Ref. (10) with permission from ELSEVIER Science).

4. Notes

1. Do not add antibiotics in the culture medium during transfection as this reduces transfection efficiency and causes cell death.
2. Gels can be purchased ready-made or produced in the laboratory (recipes can be found in laboratory handbooks). Either way, choose the percentage of your gel carefully as

this will determine the rate of migration and degree of separation between proteins>. The smaller the mass of the protein of interest, the higher the percentage of gels. The larger the mass of the protein of interest, the lower the percentage of gels.

3. 293FT cells must be passaged at least three times before using for production of lentiviral particles.
4. The health of the 293FT cells at the time of transfection has a critical effect on the success of lentivirus production. Follow the recommendations below to culture cells before use in transfection:
   a. Make sure the cells have a viability of > 90%.
   b. Subculture and maintain cells as recommended in the 293FT Cell Line manual. Do not allow cells to overgrow before passaging.
   c. Use cells that have been subcultured for less than 20 passages.
5. Cells transduced with lentiviral particles become very fragile and can be easily detached from the culture dish. Make sure cells are handled with great caution. When aspirating the medium, tilt the dish and aspirate the medium from its edge and slowly add new culture medium drop by drop against the wall of the culture dish to avoid dislodging cells and losing them.
6. Safety issues: Remember that you will be working with a medium containing an infectious virus. Follow the recommended Federal and Institutional guidelines for working with BSL-2 organisms.
7. A wide range of serial dilutions (10^{-2}–10^{-8}) may be necessary for optimal transfection. Mock controls without lentiviral transduction must be included in each experiment as a negative control.
8. Cells can be fixed by incubating the cells with 2% paraformaldehyde in PBS for 5 min. After washing the cells with PBS, resuspend the cells in PBS containing 1% FBS and store at 4°C. Use the mock-transduced cells and the lowest dilution of virus as the negative and positive controls respectively, to establish the parameters of your flow cytometer.
9. Early passage of hMSCs is recommended for lentiviral transduction. These cells can be passaged several times before they become senescent, which usually occurs after 6–7 passages.
10. To obtain optimal expression of a specific miRNA and the highest degree of target gene knockdown, care must be

exercised to achieve the optimal multiplicity of infection (MOI) to transduce the lentiviral construct into your mammalian cell line of choice. MOI is defined as the number of virus particles per cell and generally correlates with the number of integration events. Typically, miRNA expression level increases as the MOI increases.

11. Before selecting for stably transduced cells, you must first determine the minimum concentration of blasticidin required to kill your untransduced mammalian cell line (i.e., perform killing curve experiment). Typically, concentrations ranging from 2–10 μg/mL of blasticidin are sufficient to kill most untransduced cell lines.
12. Primary antibody preserved in blocking buffer containing 0.05% of sodium azide at 4°C can be used for several months.

Acknowledgments

This project was supported by grant R01 NS058780 from the National Institutes of Health, and by the Epilepsy Research Foundation through the generous support of Arlene and Arnold Goldstein Family Foundation.

References

1. Dragunow, M. (1986) Adenosine: The brain's natural anticonvulsant? *Trends Phamacol Sci* **7**, 128.
2. Boison, D. (2006) Adenosine kinase, epilepsy and stroke: Mechanisms and therapies. *Trends Pharmacol Sci* **27**, 652–658.
3. Fedele, D. E., Gouder, N., Güttinger, M., Gabernet, L., Scheurer, L., Rulicke, T., Crestani, F., and Boison, D. (2005) Astrogliosis in epilepsy leads to overexpression of adenosine kinase resulting in seizure aggravation. *Brain* **128,** 2383–2395.
4. Gouder, N., Scheurer, L., Fritschy, J.-M., and Boison, D. (2004) Overexpression of adenosine kinase in epileptic hippocampus contributes to epileptogenesis. *J Neurosci* **24,** 692–701.
5. Boison, D. (2005) Adenosine and epilepsy: From therapeutic rationale to new therapeutic strategies. *Neuroscientist* **11,** 25–36.
6. Munoz-Elias, G., Marcus, A. J., Coyne, T. M., Woodbury, D., and Black, I. B. (2004) Adult bone marrow stromal cells in the embryonic brain: Engraftment, migration, differentiation and long-term survival. *J Neurosci* **24,** 4585–4595.
7. Black, I. B., and Woodbury, D. (2001) Adult rat and human bone marrow stromal stem cells differentiate into neurons. *Blood Cells Mol Diseases* **27,** 632–636.
8. Totsugawa, T., Kobayashi, N., Okitsu, T., Noguchi, H., Watanabe, T., Matsumura, T., Maruyama, M., Fujiwara, T., Sakaguchi, M., and Tanaka, N. (2002) Lentiviral transfer of the *LacZ* gene into human endothelial cells and human bone marrow mesenchymal stem cells. *Cell Transplant* **11,** 481–488.
9. Hoelters, J., Ciccarella, M., Drechsel, M., Geissler, C., Gulkan, H., Bocker, W., Schieker, M., Jochum, M., and Neth, P. (2005) Nonviral genetic modification mediates effective transgene expression and functional RNA interference in human mesenchymal stem cells. *J Gene Med* **7,** 718–728.
10. Ren,G., Li, T., Lan, J. Q., Wilz, A., Simon, R. P., and Boison, D. (2007)

Lentiviral RNAi-induced downregulation of adenosine kinase in human mesenchymal stem cell grafts: A novel perspective for seizure control. *Experimental Neurology* **207,** 26–37.

11. Elbashir, S. M., Harborth, J., Lendeckel, W., Yalcin, A., Weber, K., and Tuschl, T. (2001a) Duplexes of 21-nucleotide RNAs mediate RNA interference in cultured mammalian cells. *Nature* **411,** 494–498.
12. Elbashir, S. M., Lendeckel, W., and Tuschl, T. (2001b) RNA interference is mediated by 21- and 22-nucleotide RNAs. *Genes Dev* **15,** 188–200.
13. Li, T., Steinbeck, J. A., Lusardi, T., Kocj, P., Lan, J. Q., Wilz. A., Segschneider, M., Simon, R. P., Brustle, O., and Boison, D. (2007) Suppression of kindling epileptogenesis by adenosine releasing stem cell-derived brain implants. *Brain* **130,** 1276–1288.
14. Franklin, K. B. J., and Paxinos, G. (1997) The mouse brain in stereotaxic coordinates. Academic, San Diego, CA.

Chapter 18

Efficient Gene Knockdowns in Mouse Embryonic Stem Cells Using MicroRNA-Based shRNAs

Jianlong Wang

Abstract

RNA interference (RNAi) is a powerful gene-knockdown technology that has been applied for functional genetic loss-of-function studies in many model eukaryotic systems, including embryonic stem cells (ESCs). Application of RNAi in ESCs allows for dissection of mechanisms by which ESCs self-renew and maintain pluripotency, and also specifying particular cell types needed for cell-replacement therapies. Potent RNAi response can be induced by expression of an microRNA-embedded short-hairpin RNA ($shRNA^{mir}$) cassette that is integrated in the genome by virus infection or site-specific recombination at a defined locus. In this chapter, I will provide detailed protocols to perform $shRNA^{mir}$-mediated RNAi studies in mouse ESCs using retrovirus infection and loxP site-directed recombination for efficient constitutive and inducible gene knockdown, respectively.

Key words: microRNA, RNA interference, microRNA-embedded short-hairpin RNA ($shRNA^{mir}$), embryonic stem cells.

1. Introduction

Embryonic stem cells (ESCs) are derived from epiblast cells within the inner cell mass of blastocysts (1, 2) and uniquely endowed with unlimited self-renewal (3) and multilineage differentiation capacity (4–6). Murine embryonic stem cells (mESCs) have become an indispensable tool for investigating genetic function both in vitro and in vivo and provided a platform to study the molecular regulation of stem-cell self-renewal and lineage commitment and cellular differentiation. The discovery of RNA interference (RNAi) (7) has provided an attractive alternative

B. Zhang, E.J. Stellwag (eds.), *RNAi and MicroRNA-Mediated Gene Regulation in Stem Cells*, Methods in Molecular Biology 650, DOI 10.1007/978-1-60761-769-3_18, © Springer Science+Business Media, LLC 2010

to traditional homology recombination-based gene knockout strategy for loss-of-function assays. Applications of RNAi in ESCs should provide valuable tools for the study of general stem-cell biology (8) as well as directed differentiation of ESCs for replacement cells/tissues in regenerative medicine (9–11). An improved understanding of pluripotency at the molecular level has led to generation of induced pluripotent stem (iPS) cells from both mouse (12, 13) and human somatic cells (14–16).

RNAi is an evolutionarily conserved, sequence-specific gene-silencing mechanism that is induced by dsRNA. MicroRNAs (miRNAs) are a class of endogenous dsRNAs that exert their effects through the RNAi pathway. Understanding the biology of the RNAi and miRNA pathways has led to the development of miRNA-based shRNA ($shRNA^{mir}$) RNAi strategy (17) that yields a higher level of siRNA and more efficient knockdown than a simple shRNA expression vector (18). The potent RNAi response and the ability to be regulated by Pol II promoters have made $shRNA^{mir}$ vectors the basis for second-generation shRNA libraries in the mouse and human genomes (17). Built on these discoveries, we have adapted the $shRNA^{mir}$-based strategy for efficient knockdown in mESCs using retrovirus transduction (19) and also developed a site-directed, virus-free, and inducible RNAi (SDVFi) system in mESCs (20). A similar nonviral, inducible RNAi approach has also been developed independently by others (21). In this chapter, I will first provide detailed procedures for a quick and easy assay to screen for a functioning $shRNA^{mir}$ cassette for a particular gene of interest (**Sections 3.1, 3.2, and 3.3**); then I will describe the method for setting up the SDVFi system in mESCs to provide a more refined and controlled experimental tool for interrogating gene function.

2. Materials

2.1. Cloning and Plasmid Preparation

1. LMPIG plasmid (*see* **Fig. 18.1**).
2. pCR2.1 TOPO cloning kit (Invitrogen).
3. pLox (ATCC cat. no. MBA-276).
4. pSalkcre.
5. EcoPak.
6. PfuUltra® PCR kit (Stratagene).
7. Rapid DNA ligation kit (Roche).
8. Plasmid miniprep and maxiprep kits (Qiagen).
9. QIAquick gel extraction kit (Qiagen).

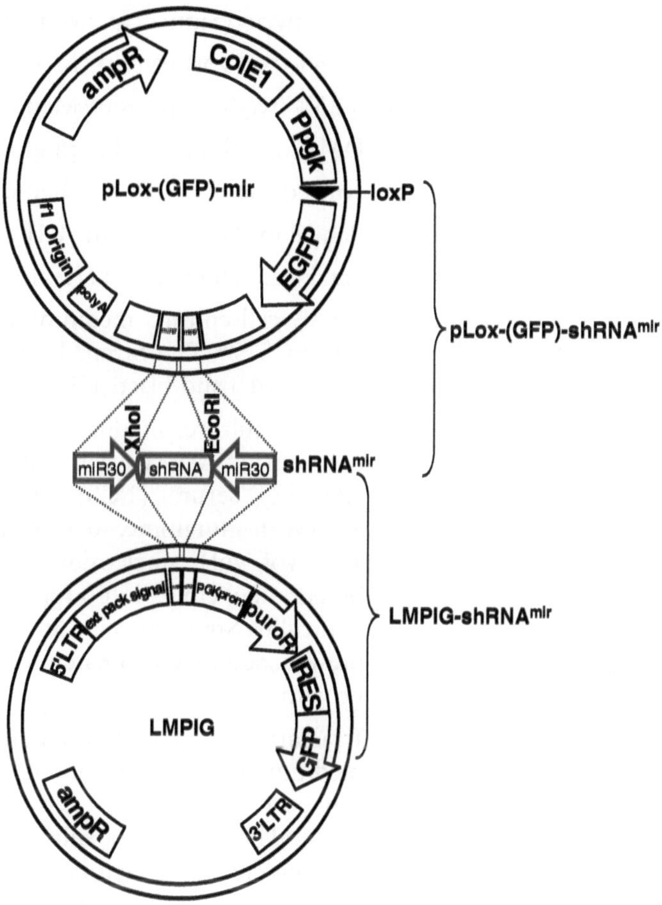

Fig. 18.1. Plasmids used in this study. (*Top*): pLox-(GFP)-shRNA^{mir} used in the SDVFI system; (*bottom*): LMPIG-shRNA^{mir} used for retrovirus-mediated, constitutive gene knockdown experiment, which also serves as a screening tool for functioning shRNAs; (*middle*) The shRNA^{mir} cassette with restriction sites *XhoI* and *EcoRI*. The shRNA can be replaced with any shRNA of the gene of interest by simple cloning into *XhoI* and *EcoRI* sites.

10. DH5α competent cells (Invitrogen).
11. Ampicillin (Sigma).

2.2. Cell Culture

1. ESGRO®/Leukemia inhibitory factor (LIF) (Chemicon, Temecula, CA).
2. Geneticin®/G418 (GIBCO cat. no. 11811): It is an aminoglycoside antibiotic that blocks polypeptide synthesis by inhibiting the elongation step in both prokaryotic and eukaryotic cells. Resistance to G418 is conferred to mammalian cells genetically engineered to express a protein product encoded by the neomycin phosphotransferase gene (i.e., neomycin-resistance gene).

3. Puromycin (Sigma cat. no. P8833).
4. 0.05% (wt/vol) trypsin (Mediatech cat. no. 25-052-CI).
5. 0.25% (wt/vol) trypsin (Mediatech cat. no. 25-053-CI).
6. DMEM with low and high glucose (Invitrogen).
7. Nucleoside mix (100X, Chemicon cat. no. ES-008-D).
8. Penicillin/Streptomycin (GIBCO cat. no. 15070-063).
9. Fetal calf serum (FCS) (Hyclone cat. no. SH30071.03).
10. Plat-E cell-culture medium: DMEM with high glucose, 10% FCS, 2% Pen-Strep, 1% L-Glutamine; add puromycin (1 μg/mL) and blasticidin (10 μg/mL) before use.
11. ES cell-culture medium: Dulbecco's Modified Eagle's Medium (DMEM) with high glucose, 15% (vol/vol) fetal calf serum (FCS)*, 0.1 mM ß-mercaptoethanol, 2 mM L-glutamine, 0.1 mM nonessential amino acid, 1% (vol/vol) nucleoside mix, 1000 U/mL recombinant leukemia inhibitory factor (LIF), 50 U/mL Penicillin/Streptomycin. Store at 4°C. *FCS needs to be prescreened batch to batch for supporting optimal ES cell growth.
12. Irradiated mouse embryonic fibroblast (iMEF) cell-culture medium: Same as ES cell-culture medium except that LIF can be omitted. MEF cells are used in the culturing of ESCs. They provide a substrate for the ESCs to grow on and secrete many factors necessary for ESCs to maintain pluripotency. Feeders are MEF cells that have been mitotically inactivated by treatment with mitomycin C or by γ-irradiation. A unique quad-resistant DR4 feeder cell line can be purchased (Open Biosystems cat. no. MES3948) or prepared from DR4 mouse embryos [JAX® mice strain STOCK Tg (DR4)1Jae/J; The Jackson Laboratory] as previously described (22).
13. Gelatin (Bacto®, DIFCO cat. no. 0143-15-1): Dissolve 5 g of gelatin in 500 mL distilled water and autoclave (1% stock). Store at room temperature indefinitely. Before use, dilute 1:10 (to make 0.1% working solution) with sterile dH_2O and filter through 0.45- μ filter apparatus.
14. Ainv15 cells (ATCC cat. no. SCRC-1029).
15. J1 ESCs (ATCC cat. no. SCRC-1010).
16. 10-cm tissue culture plate (Falcon cat. no. 35-3003).
17. 24-well plate flat-bottom (Falcon cat. no. 35-3047).
18. Culture incubator (37°C, 5% CO_2 and 100% humidity).
19. 15-mL conical tubes (Corning cat. no. 430791).
20. 50-mL conical tubes (Corning cat. no. 430829).

2.3. Plasmid Transfection and Viral Infection

1. FuGENE 6 (Roche).
2. Lipofectamine 2000 (Invitrogen).
3. Polybrene (Hexadimethrine bromide, Sigma#H9268): Make a stock solution 4 mg/mL, use it 1:1000.
4. Syringe filters (0.45 μ) (Millipore).
5. Dulbecco's Modified Eagle's Medium (DMEM) with high or low glucose (Invitrogen).

2.4. RNA Isolation and Quantitative Real-Time PCR

1. Trizol® (Invitrogen).
2. Water (Nuclease free, Ambion cat. no. am9937).
3. iCycler and SYBR Green PCR Master Mix (Biorad).
4. Primers for your gene of interest.

2.5. Total Protein Isolation and Western Blotting

1. RIPA buffer (Boston BioProducts cat. no. BP-115).
2. Protease inhibitor cocktail (Sigma cat. no. P8340).
3. Western blotting apparatus (Biorad).

2.6. FACS Sorting for GFP-Positive Cells

1. Falcon tube (#2063).
2. PBS (with calcium and magnesium) (Sigma).
3. Cell strainer (70–100 μM Nylon; BD Falcon cat. no. #352360).
4. Gentamicin (Invitrogen).

2.7. Inducible RNAi

1. Doxycycline (Sigma).
2. Genomic DNA extraction kit (Gentra Systems, Inc.).

3. Methods

3.1. Designing and Cloning shRNA Hairpins Against Your Favorite Gene

1. Go to http://codex.cshl.edu/scripts/newmain.pl Website and search for the predesigned 97-mer hairpin oligos against your favorite gene (*see* **Note 1**).
2. Order 4–5 oligos (if available) targeting different regions (5′UTR, CDS, 3′UTR) of your favorite gene (*see* **Note 2**).
3. PCR amplify the fragment using PfuUltra® PCR master mix (Stratagene) with the primers (pSM2c-Forward and pSM2c-Reverse) (*see* **Table 18.1**) and the ordered shRNA oligos (100 ng/reaction) as template. These primers will add XhoI and EcoRI sites (*see* **Table 18.1**) to the ends of your hairpin for cloning into the LMPIG vector (*see* **Section 3.2**) (see **Note 3**).

Table 18.1
Primers used in this study

Primer name	Sequence (5′→3′)
M13 reverse	CAGGAAACAGCTATGAC
T7	TAATACGACTCACTATAGGG
pSM2C forward	GATGGCTGCTCGAGAAGGTATATTGCTGTTGACAGTGAGCG
pSM2C reverse	GTCTAGAGGAATTCCGAGGCAGTAGGCA
pSM2cseq-F	GTCGACTAGGGATAACAG
pSM2cseq-R	AGTGATTTAATTTATACCA
Loxin-F	CTAGATCTCGAAGGATCTGGAG
Loxin-R	ATACTTTCTCGGCAGGAGCA

4. Clone the PCR products directly into pCR2.1 TOPO vector from Invitrogen following manufacturer's instruction (*see* **Note 4**).
5. Pick 8 white colonies for miniprep (Qiagen MiniPrep kit).
6. Verify the shRNA hairpin sequences by sequencing the miniprep DNA with primers M13 reverse and T7 (**Table 18.1**; for both + and – strands) (*see* **Note 5**).

3.2. Cloning the shRNA^{mir} Constructs

1. Digest the LMPIG vector and the miniprep TOPO plasmid (from **Section 3.1**, step 6) with *XhoI* and *EcoRI*. Excise the vector DNA band [Vector] (–8 kb) and the shRNA hairpin inserts [Insert] (–120 bp).
2. Purify digested LMPIG vector and hairpin inserts with QIAquick gel extraction kit (*see* **Note 6**).
3. Ligate the Vector and the Insert using Rapid DNA ligation kit (Roche).
4. Transform half of the ligated products into 100 μL DH5α competent cells (Invitrogen), and plate them on Luria agar ampicillin plates.
5. Inoculate 8–16 colonies for plasmid miniprep; screen the colonies by digestion of the plasmid minprep DNA with *XhoI* and *EcoRI* for correct insertion (you should see a band of –120 bp in addition to the vector band).
6. Maxiprep the positive clones. These are the LMPIG-shRNA^{mir} retroviral vectors (**Fig. 18.1**) for subsequent RNAi studies (*see* **Section 3.3**).
7. Verify the shRNA^{mir} sequence with primers pSM2cseq-F and pSM2cseq-R. (**Table 18.1**) (*see* **Note 5**).

3.3. Retroviral Delivery of shRNA^{mir} for RNAi in mESCs

3.3.1. Preparation of PLAT-E Cells and mESCs

1. Seed Plat-E cells in a 10-cm tissue culture dish with Plat-E media containing puromycin (1 μg/mL) and blasticidin (10 μg/mL) (*see* **Note 7**).
2. When they reach near 100% confluence, split 1:5 onto new tissue culture dishes. Depending on the number of samples, another round of expansion of Plat-E cells may be needed.
3. On the day of transfection, Plat-E cells should reach 60–70% confluence. Replace with fresh medium without puromycin and blasticidin immediately before transfection.
4. A day or days prior to transfection of Plat-E cells, thaw out J1 or mESCs of your choice and seed them on top of a layer of irradiated mouse embryonic fibroblasts (iMEFs) in a 10-cm dish. The iMEFs (-1×10^6 cells) should be seeded a few hours or a day before thawing mESCs (*see* **Note 8**).

3.3.2. Transfection of Plat-E Cells with LMPIG-shRNA^{mir} Plasmids

1. Prepare plasmids: Mix 20 μg DNA (10 μg LMPIG-shRNA^{mir} plasmid + 10 μg Ecopak); leave them at room temperature (*see* **Note 7**).
2. Prepare diluted FuGENE 6: Per transfection, add 1 mL of DMEM with low glucose to a 15-mL Falcon tube, then add 60 μL FuGENE 6 reagent directly to medium in tubes (avoid contacting wall of the tube). Incubate at RT for 5 min (*see* **Note 9**).
3. Add 1 mL of the diluted FuGENE 6 (Step 2) into the DNA mix (Step 1). Incubate at RT for 15 min.
4. Add mixed DNAFuGENE 6 complex dropwise onto the Plat-E cells (from **Section 3.3.1**, step 3). Return to incubator and culture O/N.
5. After 24 h, gently aspirate the medium and add 5 mL of ES medium. Return to incubator and culture O/N.
6. Split ESCs in a ratio such that the number of ESC plates should be equal to or larger than that of Plat-E cells (*see* **Section 3.3.1**, step 2) and ESCs should reach 60–70% confluence by the next day for use in **Section 3.3.3**, step 3.

3.3.3. Infection of mESCs with Retroviruses

1. Collect the medium from dishes (in **Section 3.3.2**, step 5) after 24 h (designated 1° virus supernatant) and add another 5 mL of fresh ES medium. Return plates to incubator.
2. Dilute 1° virus supernatant 1:2 by adding 5 mL of ES medium (total 10 mL), filtered with a 0.45-μm syringe filter (*see* **Note 10**).

3. Gently aspirate the medium from ESCs (from **Section 3.3.2**, step 6) and add the 10 mL of filtered virus-containing medium to ESCs.
4. Add 10 μL polybrene (final 4 μg/mL) to each dish. Return to incubator.
5. On second day, collect the 5 mL virus medium (designated 2° virus supernatant) from Plat-E cells (in **Section 3.3.3**, step 1); rinse each dish with another 5 mL of ES medium. Pool total 10 mL ES medium and filter with a 0.45-μm syringe filter.
6. Gently aspirate the old medium from ESCs and add the 2° supernatant (i.e., 10 mL filtered virus-containing media) to ESCs. The cells should now reach near 90% confluence.
7. Add 10 μL of polybrene (final 4 μg/mL) to each dish. Return to incubator and culture O/N.
8. Replace old medium with fresh medium containing 1–2 μg/mL puromycin daily for the next few days (*see* **Note 11**).

3.3.4. Verification of Gene Knockdown

3.3.4.1. Harvest Cells Directly for RNA and Protein Extractions

1. On day 3 (i.e., 48 hr after initiation of puromycin selection) or later, remove ES medium by aspiration, and rinse ESCs once with 0.05% trypsin. Then, add a sufficient amount of 0.25% trypsin to cover the ESCs and incubate at 37°C for 35 min. The ESCs become detached from the vessels and can be collected after neutralization of the trypsin with 3 × vol of ES cell medium (pipetting up and down to mix them).
2. Split cell suspension into halves, one for RNA, the other for total protein. Centrifuge at 200 × *g* for 5 min to harvest cells.
3. Resuspend the cell pellets with 10 mL of PBS and centrifuge at 200 × *g* for 5 min.
4. Resuspend the cell pellets in Trizol for RNA extraction, and in RIPA buffer for total protein extraction.
5. Perform standard Q real-time PCR and Western blotting to verify gene knockdown (*see* **Note 12**).

3.3.4.2. Harvest Cells by Sorting GFP-Positive Cells for RNA and Protein Extractions

1. On day 3 (i.e., 48 h after initiation of puromycin selection) or later, remove ES medium by aspiration, and rinse ESCs once with 0.05% trypsin. Then, add a sufficient amount of 0.25% trypsin to cover the ESCs and incubate at 37°C for 3–5 min. The ESCs become detached from the vessels and can be collected after neutralization of the trypsin with

3 × vol of ES medium (pipetting up and down to mix them) (*see* **Note 13**).

2. Centrifuge at 200 × *g* for 5 min to harvest cells.
3. Aspirate and resuspend cell pellets in 10 mL of PBS (with calcium and magnesium) containing 5% FCS. Wash twice with 10 mL of the same buffer.
4. Resuspend cell pellets with appropriate amount of the same buffer to make 2–5 × 10^6/mL cells for sorting.
5. Filter cell suspension with 70–100 μm cell strainer to remove cell clumps. Have collection tubes prefilled with −1 mL of 100% heat-inactivated FCS containing Pen/Strep (50 U/mL), gentamicin (50 μg/mL), and puromycin (1–2 μg/mL) (*see* **Note 14**).
6. Sort cells for high, medium, and low GFP intensity. Collect cells in a Falcon tube prepared in Step 5.
7. After sorting, you can either harvest all the cells for RNA or total protein extraction following **Section 3.3.4.1**, step 4. (if you have enough cells) or reculture them for expansion or single-colony formation (*see* next step).
8. Add directly 5–10 mL of ES medium to the sorted cells and transfer them to a 10-cm iMEF dish or 6-well iMEF plate depending on cell numbers after sorting. Return to incubator and culture for a few more days under 37°C/5% CO_2 (*see* **Note 15**).
9. Perform RNA and protein preparation for real-time PCR and Western blotting for gene knockdown at RNA and protein levels, respectively.

3.4. Site-Directed, Virus-Free, and Inducible Expression of shRNA^{mir} for RNAi in mESCs

3.4.1. Establishment of ESCs Expressing shRNA^{mir} in a Defined Locus

1. Digest the plasmid carrying the validated shRNA^{mir} (in **Section 3.3.4**) with *XhoI* and *EcoRI* to release the shRNA and clone it into the pLox-mir or pLox-GFP-mir (to mark the mir expression with GFP reporter) vector previously digested with *XhoI* and *EcoRI*. The resultant vector is pLox-(GFP)-shRNA^{mir} (**Fig. 18.1**) (*see* **Note 16**).
2. Make miniprep and maxiprep of pLox-(GFP)-shRNA^{mir}, and verify the plasmids by restriction digests and direct DNA sequencing.

3. Thaw and grow Ainv15 ESCs on a 6-well iMEF plate. When ESCs reach a density necessary for passage, split a single well into five to six similar wells. Incubate overnight or until cells reach 80–90% confluence.
4. Prepare master mixes in an Eppendorf tube or a 15-mL Falcon tube per sample: Add 10 μL Lipofectamine 2000 into 250 μL serum-free DMEM with high glucose (*see* **Note 17**).
5. In a 1.5-mL Eppendorf tube, add 5 μg of pLox-(GFP)-shRNA^{mir} and 5 μg of pSalkcre DNA into 250 μL serum-free medium.
6. Incubate the tubes from Steps 4 and 5 for 5 min at room temperature.
7. Aliquot 250 μL of Lipofectamine 2000 mix (Step 4) into 250 μL Eppendorf tubes containing DNA mix (Step 5), and mix gently by inverting the tubes. Incubate this mixture for 10 min at RT to allow the lipid and DNA complexes to form.
8. While the incubation is proceeding, aspirate ES cell medium from Ainv15 cells, wash each well with PBS, treat cells using trypsin, and harvest by centrifugation at 400 × *g* for 5 min in 15-mL Falcon tubes.
9. Resuspend cell pellet with DNA/lipid complexes (Step 7) and incubate at RT for 10 min.
10. Add DNA/lipid complexes/cells (from Step 9) dropwise to gelatin-coated wells of a 6-well plate, rock the plate back and forth to distribute the cell suspension, and incubate overnight under 37°C/5% CO_2 in tissue culture incubator. Prepare same number of 10-cm iMEF dishes for use the next day.
11. On the second day, the cells in the 6-well plate should be near 100% confluence. Trypsinize cells, harvest by centrifugation as in step 8, resuspend in 10 mL of fresh ES medium, transfer cells into 10-cm iMEF plates, and incubate under 37°C/5% CO_2.
12. On the fourth day (48 h later), add 10 μL of G418 (final 300 μg/mL) directly to each dish to start drug selection.
13. Replace the old medium daily with fresh medium containing 300 μg/mL G418 for the next 10–15 days (*see* **Note 18**). Pick the emerging G418-resistant clones, expand cells for genomic DNA extraction (*see* Gentra manuals), and freeze cell stocks in 90% fetal bovine serum/10% DMSO (designated Ainv/pLox-(GFP)-shRNA^{mir}).
14. PCR confirm the correct targeting in positive clones using primers Loxin-F/Loxin-R (*see* **Table 18.1**). PCR

amplification products are analyzed by agarose gel electrophoresis, and the presence of a –500-bp band that is amplified across the loxP site indicates a correct targeting event.

3.4.2. Inducible Gene Knockdown in Ainv/pLox-(GFP)-shRNA^mir Cells

1. Thaw Ainv/pLox-(GFP)-shRNA^{mir} cells and expand them onto 2 wells of a 24-well iMEF plate.
2. Grow cells until they reach 40–50% confluence.
3. Add doxycycline (1–2 μg/mL) to one of the wells.
4. Replace old medium with fresh medium daily (adding 1–2 μg/mL doxycycline to the well under selection) for the next 3–7 days. Split cells into several new wells of a 24-well iMEF plate when cells are near confluence.
5. Harvest cells from one well for RNA extraction, and another well of cells for total protein lysates as described in **Section 3.3.4.1**.
6. Perform quantitative real-time PCR and/or Western blotting to validate knockdown of the gene of interest at RNA and protein levels, respectively (*see* **Note 19**).

4. Notes

1. RNAi Codex provides a single database that curates publicly available RNAi resources including the Hannon-Elledge shRNA libraries (mouse and human) that are available through Open Biosystems. The Codex provides the most complete access to this growing resource, allowing investigators to access available clones and clones that are soon to be released. Independent of the optimal shRNA design by the RNAi Codex, control experiments are necessary to confirm the specificity of an RNAi phenotype (*see* **Note 2**).
2. If a pSM2-shRNA already exists in the Open Biosystems collection (http://www.openbiosystems.com/expression_arrest_shrna_libraries.php), and/or the oligos in http://codex.cshl.edu/scripts/newmain.pl search are marked as "released", you can order them from Open Biosystems as a bacterial stock. As mentioned in **Note 1**, control shRNAs should also be ordered and processed simultaneously. Open Biosystems offers pSM2 Retroviral shRNA^{mir} controls in glycerol stock (Catalog # RHS1705, RHS1706, RHS1707). The Firefly Luciferase pSM2 shRNA^{mir} is a positive control designed against pGL3 Firefly Luciferase

(Promega, cat. no. E1741). The eGFP pSM2 shRNA^{mir} is a positive control designed against the enhanced GFP reporter (Invitrogen, cat. no. v355-20; GenBank accession number: pEGFP U76561). The nonsilencing pSM2 shRNA^{mir} is a negative control containing a target sequence that does not match any known mammalian genes. Depending on your experimental setting, the Firefly Luciferase shRNA^{mir} and the eGFP shRNA^{mir} can also be used as negative controls against your gene of interest.

3. High-fidelity amplification PCR kit (e.g., Stratagene's PfuUltra®, Roche's High-Fidelity®, and NEB's Phusion®) is preferred for amplification of shRNA hairpin sequences. In any case, when using 100 ng/reaction of the oligo as template, 12–16 cycles are sufficient to generate enough PCR product for cloning. Over-amplification will increase errors in the final products.
4. TOPO cloning of PCR products amplified with high-fidelity polymerase such as PfuUltra® requires pretreatment of the PCR products with Taq polymerase to add T/A overhangs (*see* TOPO cloning kit for details).
5. When setting up sequencing reactions, you should include 5% (vol/vol) dimethyl sulfoxide (DMSO) to resolve certain compressions caused by strong shRNA^{mir} hairpin secondary structure.
6. When transferring *XhoI/EcoRI* hairpin fragments from one vector to another using a gel-purification step, "melt" agarose at lower temperature (42 instead of 50 degrees) to reduce probability of melting hairpins (resulting in snap back ss DNA).
7. A potent retrovirus packaging cell line named Platinum-E (Plat-E) was generated based on the 293T cell line. Plat-E cells have been engineered to stably express the *gag-pol* and *env* genes under the strong EF1α promoter for efficient virus packaging. The high titer of retroviruses derived from the Plat-E cells can be maintained by simply culturing the cells in the presence of selection drugs (puromycin and blasticidin) (23). EcoPak is a plasmid that expresses the *gag-pol-env* packaging functions as previously described by another group (24). The use of EcoPak in transfection of Plat-E cells is optional.
8. This step should be carefully planned so that you will have enough ESCs for infection (*see* **Section 3.3.3**) after splitting when they reach confluence (*see* **Section 3.3.2**, step 6).
9. If not using EcoPak, then mix 10 μg shRNA plasmid with 30 μL FuGENE 6 (dilute with 500 μL DMEM

with low glucose). FuGENE 6 as well as other lipid-based transfection reagent (e.g., Lipofectamine 2000) should be warmed up to ambient temperature (approximately 10–15 min at room temperature) prior to use and added directly to the media while avoiding contact with the walls of tubes/plastics. Chemical residues in plastic vials can significantly decrease the biological activity of the reagent. In addition, no drug should be present in DMEM (low glucose) during FuGENE 6 or Lipofectamine 2000 dilution.

10. The virus supernatant should be filtered with a 0.45-μm syringe filter (recommended). However, if there are obvious big cell clumps floating in the supernatant, you should centrifuge the supernatant at 1500 rpm for 5 min to remove clumped cells before using the syringe filter, otherwise the syringe filter will be clogged by clumped cells.
11. The puromycin resistance is conferred on the infected cells by expression of the puromycin-ires-GFP expression cassette in the LMPIG-shRNA^{mir} vector (*see* **Fig. 18.1**). Selection of puromycin-resistant cells will enrich the infected cell population for downstream analyses.
12. Based on the variable expression of GFP dependent fluorescence, we concluded that puromycin-selected cell populations are represented by cells that can be divided into those with low, moderate, and high levels of gene knockdown. Moreover, when genes important for stem-cell self-renewal were studied, we observed that the low-level knockdown cells have a growth advantage over the high-level knockdown cells, which leads them to outcompete the high-level knockdown cells. Consequently, prolonged puromycin selection should be avoided. A time course extending over a 3-day period from day 1 to day 3 (within 72 h post-puromycin selection) during which the ratios of low, moderate, and highly fluorescent cells are monitored may prove useful in estimating dynamic changes in the cell population. As an alternative procedure, **Section 3.3.4.2** provides a solution to this problem by sorting out the GFP high, medium, and low populations for downstream analyses.
13. Depending on the infection efficiency, the number of puromycin-resistant ESCs after several days of puromycin selection may not be enough for downstream RNA and protein analyses after GFP sorting. Therefore, if more cells are deemed necessary for downstream analyses, ESCs should be further expanded and grown onto 10-cm iMEF plates, provided that knockdown cells have not differentiated.

14. Pen/Strep, gentamicin, and puromycin were added to the collection tube to prevent potential contamination from bacteria and other irrelevant cells from sorting. This is only necessary if the sorted cells will be further cultured.
15. If a single cell colony is preferred, then collect 400–800 GFP+ cells, and culture them in a 10-cm iMEF dish to allow individual clones to form. Many cells will die due to the sorting stress, but enough cells survive and will grow out as single colonies after 10–14 days, provided that the knockdown of the gene does not result in cell death or differentiation.
16. The advantage of using pLox-GFP-shRNA^{mir} over pLox-shRNA^{mir} is that the expression of shRNA^{mir} will be marked by the GFP reporter upon doxycycline induction. In addition, it has been reported that an increased spacer between the promoter and the shRNA^{mir} cassette delivered by lentiviral vectors may enhance the knockdown levels (25), although the mechanism is unclear.
17. Transfection of Ainv15 cells with pLox and pSalkcre constructs has only been reported using electroporation (26), which often yields only a few G418-resistant homology recombinants. In this chapter, I have tested using Lipofectamine 2000 as transfection reagent to introduce pLox-GFP-shRNA^{mir} and pSalkcre into Ainv15 cells, and found that Lipofectamine 2000 transfection yields almost twice as many G418-resistant colonies than electroporation.
18. Most cells die after 3–4 days of selection using 300–350 μg/mL G418 in complete ES cell growth media. Culture and change the drug-containing medium daily until colonies appear around day 10–14.
19. Due to the single copy integration of the shRNA^{mir} in the SDVFi system, the knockdown of high-abundance genes may not be as efficient as that in the retroviral system, which often involves multicopy shRNA^{mir} random integration into the genome. In this case, a tandem, multiple shRNA^{mir}-cassette strategy may be attempted.

Acknowledgments

The author would like to thank Dr. Stuart H. Orkin, an Investigator of Howard Hughes Medical Institute, for his support of the author's postdoctoral training in his lab when the method was initially developed. The author's current work is supported by the Seed Fund from the Black Family Stem Cell Institute in Mount Sinai School of Medicine.

References

1. Evans, M. J., and Kaufman, M. H. (1981) Establishment in culture of pluripotential cells from mouse embryos. *Nature* **292**, 154–156.
2. Martin, G. R. (1981) Isolation of a pluripotent cell line from early mouse embryos cultured in medium conditioned by teratocarcinoma stem cells. *Proc Natl Acad Sci USA* **78**, 7634–7638.
3. Chambers, I., and Smith, A. (2004) Self-renewal of teratocarcinoma and embryonic stem cells. *Oncogene* **23**, 7150–7160.
4. Spagnoli, F. M., and Hemmati-Brivanlou, A. (2006) Guiding embryonic stem cells towards differentiation: Lessons from molecular embryology. *Curr Opin Genetics Dev* **16**, 469–475.
5. Odorico, J. S., Kaufman, D. S., and Thomson, J. A. (2001) Multilineage differentiation from human embryonic stem cell lines. *Stem Cells* 19, 193–204.
6. Gadue, P., Huber, T. L., Nostro, M. C., Kattman, S., and Keller, G. M. (2005) Germ layer induction from embryonic stem cells. *Exp Hematol* **33**, 955–964.
7. Fire, A., Xu, S., Montgomery, M. K., Kostas, S. A., Driver, S. E., and Mello, C. C. (1998) Potent and specific genetic interference by double-stranded RNA in *Caenorhabditis elegans. Nature* **391**, 806–811.
8. Ivanova, N., Dobrin, R., Lu, R., Kotenko, I., Levorse, J., DeCoste, C., Schafer, X., Lun, Y., and Lemischka, I. R. (2006) Dissecting self-renewal in stem cells with RNA interference. *Nature* **442**, 533–538.
9. Ding, L., and Buchholz, F. (2006) RNAi in embryonic stem cells. *Stem Cell Rev* **2**, 11–18.
10. Heidersbach, A., Gaspar-Maia, A., McManus, M. T., and Ramalho-Santos, M. (2006) RNA interference in embryonic stem cells and the prospects for future therapies. *Gene Ther* **13**, 478–486.
11. Spankuch, B., and Strebhardt, K. (2005) RNA interference-based gene silencing in mice: The development of a novel therapeutical strategy. *Curr Pharm Des* **11**, 3405–3419.
12. Okita, K., Ichisaka, T., and Yamanaka, S. (2007) Generation of germline-competent induced pluripotent stem cells. *Nature* **448**, 313–317.
13. Takahashi, K., and Yamanaka, S. (2006) Induction of pluripotent stem cells from mouse embryonic and adult fibroblast cultures by defined factors. *Cell* **126**, 663–676.
14. Park, I. H., Zhao, R., West, J. A., Yabuuchi, A., Huo, H., Ince, T. A., Lerou, P. H., Lensch, M. W., and Daley, G. Q. (2008) Reprogramming of human somatic cells to pluripotency with defined factors. *Nature* **451**, 141–146.
15. Takahashi, K., Tanabe, K., Ohnuki, M., Narita, M., Ichisaka, T., Tomoda, K., and Yamanaka, S. (2007) Induction of pluripotent stem cells from adult human fibroblasts by defined factors. *Cell* **131**, 861–872.
16. Yu, J., Vodyanik, M. A., Smuga-Otto, K., Antosiewicz-Bourget, J., Frane, J. L., Tian, S., Nie, J., Jonsdottir, G. A., Ruotti, V., Stewart, R., Slukvin, II, and Thomson, J. A. (2007) Induced pluripotent stem cell lines derived from human somatic cells. *Science* **318**, 1917–1920.
17. Chang, K., Elledge, S. J., and Hannon, G. J. (2006) Lessons from Nature: microrna-based shRNA libraries. *Nat Methods* **3**, 707–714.
18. Silva, J. M., Li, M. Z., Chang, K., Ge, W., Golding, M. C., Rickles, R. J., Siolas, D., Hu, G., Paddison, P. J., Schlabach, M. R., Sheth, N., Bradshaw, J., Burchard, J., Kulkarni, A., Cavet, G., Sachidanandam, R., McCombie, W. R., Cleary, M. A., Elledge, S. J., and Hannon, G. J. (2005) Second-generation shRNA libraries covering the mouse and human genomes. *Nat Genet* **37**, 1281–1288.
19. Wang, J., Rao, S., Chu, J., Shen, X., Levasseur, D. N., Theunissen, T. W., and Orkin, S. H. (2006) A protein interaction network for pluripotency of embryonic stem cells. *Nature* **444**, 364–368.
20. Wang, J., Theunissen, T. W., and Orkin, S. H. (2007) Site-directed, virus-free, and inducible RNAi in embryonic stem cells. *Proc Natl Acad Sci USA* **104**, 20850–20855.
21. Lohmann, F., and Bieker, J. J. (2008) Activation of Eklf expression during hematopoiesis by Gata2 and Smad5 prior to erythroid commitment. *Development (Cambridge, England)* **135**, 2071–2082.
22. Conner, D. A. (2001) Mouse embryo fibroblast (MEF) feeder cell preparation,

In: Ausubel, F. M. et al. (ed) *Curr Protoc Mol Biol* Chapter 23, Unit 23.2.1–23.2.7.
23. Morita, S., Kojima, T., and Kitamura, T. (2000) Plat-E: An efficient and stable system for transient packaging of retroviruses. *Gene Ther* 7, 1063–1066.
24. Gavrilescu, L. C., and Van Etten, R. A. (2007) Production of replication-defective retrovirus by transient transfection of 293T cells. *J Vis Exp* **10**, 550.
25. Stegmeier, F., Hu, G., Rickles, R. J., Hannon, G. J., and Elledge, S. J. (2005) A lentiviral microRNA-based system for single-copy polymerase II-regulated RNA interference in mammalian cells. *Proc Natl Acad Sci USA* **102**, 13212–13217.
26. Kyba, M., Perlingeiro, R. C., and Daley, G. Q. (2002) HoxB4 confers definitive lymphoid-myeloid engraftment potential on embryonic stem cell and yolk sac hematopoietic progenitors. *Cell* **109**, 29–37.

Subject Index

B. Zhang, E.J. Stellwag (eds.), *RNAi and MicroRNA-Mediated Gene Regulation in Stem Cells*, Methods in Molecular Biology 650, DOI 10.1007/978-1-60761-769-3, © Springer Science+Business Media, LLC 2010

I

K

L

M

N

O

P

U

V

W

X

MIX
Papier aus verantwortungsvollen Quellen
Paper from responsible sources
FSC® C105338

If you have any concerns about our products,
you can contact us on
ProductSafety@springernature.com

In case Publisher is established outside the EU,
the EU authorized representative is:
Springer Nature Customer Service Center GmbH
Europaplatz 3, 69115 Heidelberg, Germany

Printed by Libri Plureos GmbH
in Hamburg, Germany